I0056529

Metallurgy Fundamentals

Metallurgy Fundamentals

Dakota Owen

NY RESEARCH
P R E S S

New York

Published by NY Research Press
118-35 Queens Blvd., Suite 400,
Forest Hills, NY 11375, USA
www.nyresearchpress.com

Metallurgy Fundamentals
Dakota Owen

© 2020 NY Research Press

International Standard Book Number: 978-1-63238-798-1 (Hardback)

This book contains information obtained from authentic and highly regarded sources. All chapters are published with permission under the Creative Commons Attribution Share Alike License or equivalent. A wide variety of references are listed. Permissions and sources are indicated; for detailed attributions, please refer to the permissions page. Reasonable efforts have been made to publish reliable data and information, but the authors, editors and publisher cannot assume any responsibility for the validity of all materials or the consequences of their use.

Trademark Notice: Registered trademark of products or corporate names are used only for explanation and identification without intent to infringe.

Cataloging-in-Publication Data

Metallurgy fundamentals / Dakota Owen.
    p. cm.
Includes bibliographical references and index.
ISBN 978-1-63238-798-1
1. Metallurgy. 2. Iron--Metallurgy. 3. Steel--Metallurgy.
4. Nonferrous metals--Metallurgy. I. Owen, Dakota.
TN673 .M48 2020
669--dc23

TABLE OF CONTENTS

This book is a culmination of my many years of practice in this field. I attribute the success of this book to my support group. I would like to thank my parents who have showered me with unconditional love and support and my peers and professors for their constant guidance.

Metallurgy deals with the study of the chemical and physical behavior of metallic elements, their inter-metallic compounds as well as their alloys. It falls under the domain of materials science and engineering. It is used for the separation of metals from their ore. Metallurgy also deals with the application of science to the production of metals and the engineering of metal components. Metallurgy is broadly divided into ferrous metallurgy and non-ferrous metallurgy. Ferrous metallurgy includes the processes and alloys that contain iron. Non-ferrous metallurgy deals with the processes involving metals and alloys such as aluminium, copper, lead, brass, etc. This textbook presents the complex subject of metallurgy in the most comprehensible and easy to understand language. Most of the topics introduced in this book cover new techniques and the applications of metallurgy. It will provide comprehensive knowledge to the readers.

The details of chapters are provided below for a progressive learning:

Chapter – Introduction

Metallurgy combines the principles of material science and engineering to study the chemical and physical behavior of metallic and inter-metallic compounds and alloys. Ferrous metallurgy, metallurgy of iron and steel, steelmaking, etc. are some of its aspects. These diverse aspects of metallurgy have been thoroughly discussed in this chapter.

Chapter – Extractive Metallurgy

The science of extracting valuable metals from their ore and refining them to achieve their purest form is referred to as extractive metallurgy. Mineral processing, smelting, hydrometallurgy, pyrometallurgy, electrometallurgy, etc. are some the concepts related to it. This chapter discusses in detail the concepts related to extractive metallurgy.

Chapter – Physical Metallurgy

Physical metallurgy is concerned with the physical properties of metals and alloys. Some of the properties which are studied within this field are mechanical, magnetic and thermal properties. This chapter has been carefully written to provide an easy understanding of the various properties and applications of physical metallurgy.

Chapter – Powder Metallurgy

Powder metallurgy is the science of creating various materials and compounds from metal powder. It includes the processes such as powder production, mixing powders for PM processing, sintering, etc. The chapter closely examines these processes and properties of powder metallurgy to provide an extensive understanding of the subject.

Chapter – Metalworking

The process of working with metals to create individual parts or large structures is known as metalworking. Some of its components consist metal forming, molding, extrusion, metal casting, metal cutting, metal spinning, etc. This chapter delves into study of these components to provide in-depth knowledge for the subject.

Dakota Owen

Introduction

Metallurgy combines the principles of material science and engineering to study the chemical and physical behavior of metallic and inter-metallic compounds and alloys. Ferrous metallurgy, metallurgy of iron and steel, steelmaking, etc. are some of its aspects. These diverse aspects of metallurgy have been thoroughly discussed in this chapter.

METALS

With the exception of hydrogen, all elements that form positive ions by losing electrons during chemical reactions are called metals. Thus metals are electropositive elements with relatively low ionization energies. They are characterized by bright luster, hardness, ability to resonate sound and are excellent conductors of heat and electricity. Metals are solids under normal conditions except for Mercury.

All the things around us are made of 100 or so elements. These elements were classified by Lavoisier in to metals and non-metals by studying their properties. The metals and non-metals differ in their properties.

Main Group Al, Ga, In, Sn, Tl, Pb, Bi, Po.

Alkali elements Li, Na, K, Rb, Cs, Fr.

Alkaline earth elements Be, Mg, Ca, Sr, Ba.

General Physical Properties of the Metals

The metals have a shiny appearance, they show a metallic luster. Due to their shiny appearance they can be used in jewellery and decorations. Particularly gold and silver are widely used for jewellery. In the old days, mirrors were made of shiny metals like silver. Silver is a very good reflector. It reflects about 90% of the light falling on it. All modern mirrors contain a thin coating of metals.

Metals are mostly harder to cut. Their hardness varies from one metal to another. Some metals like sodium, potassium and magnesium are easy to cut.

Metals on being hammered can be beaten into thinner sheets. This property is called Malleability. Most metals are malleable. Gold and Silver metals are the most malleable metals. They can be hammered into very fine sheets. Thin aluminium foils are widely used for safe wrapping of medicines, chocolates and food material.

Wires are made from copper, aluminium, iron and magnesium. This property of drawing the metal in to thin wires is called ductility. Most metals are ductile.

Electric wires in our homes are made of aluminium and copper. They are good conductor of electricity. Electricity flows most easily through gold, silver, copper and aluminium. Gold and silver are used for fne electrical contacts in computers. Copper wires are used in electrical appliances while aluminium is cheaper is generally used for making electrical cables.

Cooking utensils and water boilers are also made of iron, copper and aluminium, because they are good conductors of heat.

Metals are generall sonorous. That is they make a ringing sound when struck. Therefore, they are used for making bells. Metal wires are used in musical instruments.

Metals such as iron are very strong. Therefore, it is therefore, widely used in the construction of buildings, bridges, railway lines, carriages, vehicles and machinery.

All metals except Mercury, exist in the solid form at room temperature. Therefore, they retain their shapes under normal conditions.

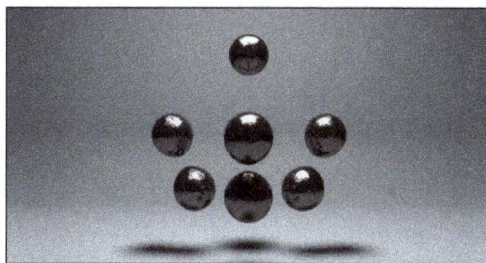

Metals have high melting points.

Metals have high tensile strength, that is they can be stretched to some degree without breaking. Metals like tungsten has high tensile strength.

No two metals are absolutely identical. For example, iron is magnetic and copper is not.

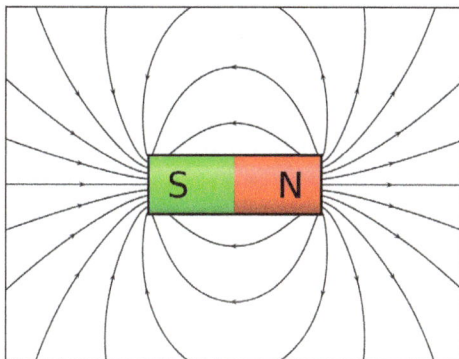

Gold an Platinum are malleable and ductile but do not react with water.Sodium is highly reactive and reacts vigorously with water to form a solution of sodium hydroxide.

Chemical Properties of Metals

Metals are also called electropositive elements because the metal atoms form positively charged ion by losing electrons. Following are the important chemical reactions of metals which takes place due to the electropositive character of metals.

Reaction of Metals with Oxygen

Almost all metals react with oxygen to form metal oxides. But different metals react with oxygen at different intensities. For example, sodium metal is always kept immersed in kerosene oil. Because, if we keep it open, it reacts so vigorously with oxygen present in air that it catches fire. We have already discussed that the oxides of metals are basic in nature. As all the metals have different reactivity so they combine with oxygen at different temparature.

1. Sodium metal reacts with oxygen of air at room temperature to form basic sodium oxide.

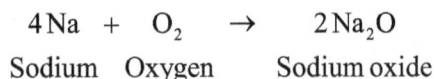

$$4\,Na \ + \ O_2 \quad \rightarrow \quad 2\,Na_2O$$
Sodium   Oxygen   Sodium oxide

2. On heating, magnesium metal burns in air giving magnesium oxide.

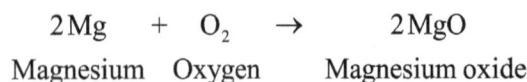

$$2\,Mg \quad + \quad O_2 \quad \rightarrow \quad 2\,MgO$$
Magnesium   Oxygen   Magnesium oxide

3. Zinc metal burns in air only on strong heating to form zinc oxide.

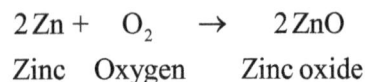

$$2\,Zn + \quad O_2 \quad \rightarrow \quad 2\,ZnO$$
Zinc   Oxygen   Zinc oxide

Generally, metal oxides are insoluble in water. But some metal oxides are able to dissolve in water to form metal hydroxides (or alkali). For example, oxides of sodium and potassium dissolve in water to form sodium hydroxide and potassium hydroxide respectively.

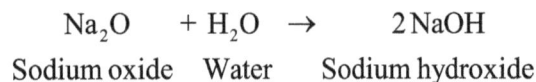

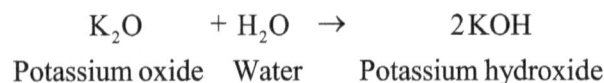

$$Na_2O \quad + H_2O \quad \rightarrow \quad 2\,NaOH$$
Sodium oxide   Water   Sodium hydroxide

$$K_2O \quad + H_2O \quad \rightarrow \quad 2\,KOH$$
Potassium oxide   Water   Potassium hydroxide

In the same way sulphur reacts with oxygen of air to form acidic sulphur dioxide.

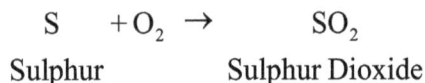

$$S \quad + O_2 \rightarrow \quad SO_2$$
Sulphur   Sulphur Dioxide

Reaction of Metals with Water

Metals react with water to produce metal oxide (or metal hydroxide) and hydrogen gas. But, all metals do not react with water at equal intensity. The metals which are very reactive can react even with cold water while the other metals react with hot water or with steam. For example:

1. Sodium, potassium and calcium metal can react with cold water to produce their hydroxides and hydrogen gas.

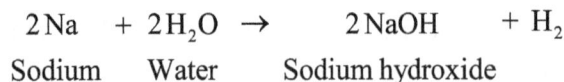

$$2\,Na \quad + 2\,H_2O \quad \rightarrow \quad 2\,NaOH \quad + H_2$$
Sodium   Water   Sodium hydroxide

$$2K \quad + 2H_2O \rightarrow \qquad 2KOH \qquad + H_2$$
Potassium   Water   Potassium hydroxide

$$Ca \quad + 2H_2O \rightarrow \qquad Ca(OH)_2 \quad + H_2$$
Calcium   Water   Calcium hydroxide

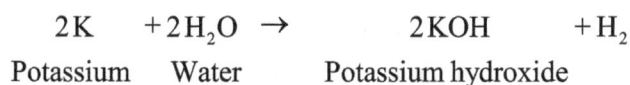

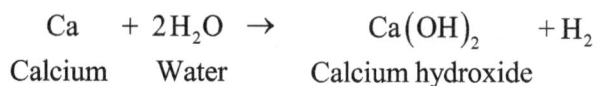

2. Magnesium, zinc and iron react with hot water to produce metal oxide and hydrogen gas.

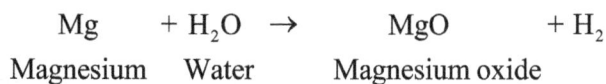

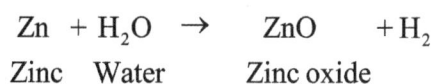

$$Mg \quad + H_2O \rightarrow \qquad MgO \qquad + H_2$$
Magnesium   Water   Magnesium oxide

$$Zn + H_2O \rightarrow \quad ZnO \quad + H_2$$
Zinc   Water   Zinc oxide

$$3Fe + 4H_2O \rightarrow \quad Fe_3O_4 \quad + H_2$$
Iron   Water   Iron oxide

Reaction of Metals with Dilute Acids

When a metal reacts with a dilute acid then a metal salt and hydrogen gas are formed. For example: Sodium, magnesium and zinc reacts with dilute hydrochloric acid to form their salts and hydrogen gas.

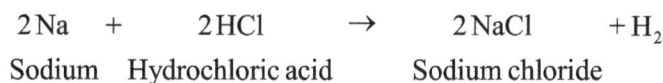

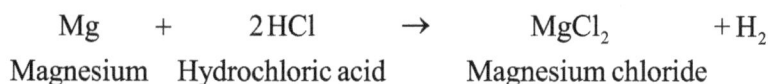

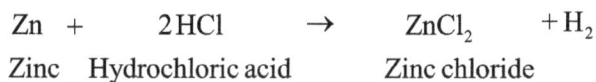

$$2Na \quad + \qquad 2HCl \qquad \rightarrow \qquad 2NaCl \qquad + H_2$$
Sodium   Hydrochloric acid   Sodium chloride

$$Mg \quad + \qquad 2HCl \qquad \rightarrow \qquad MgCl_2 \qquad + H_2$$
Magnesium   Hydrochloric acid   Magnesium chloride

$$Zn + \qquad 2HCl \qquad \rightarrow \quad ZnCl_2 \quad + H_2$$
Zinc   Hydrochloric acid   Zinc chloride

Reaction of Metals with Salt Solutions

If a more reactive metal is put in the salt solution of a less reactive metal, the more reactive metal displaces the less reactive metal from its salt solution. These reactions are called displacement reaction.

Reaction of Copper with Silver Nitrate Solution

If a piece of copper metal is placed in colourless solution of silver nitrate for some time, the colour of the solution becomes blue and a shining white deposit of silver metal is formed on the piece of copper. Actually, in this reaction copper metal is more reactive than silver present in silver nitrate solution. So, copper displaces silver from silver nitrate solution to form copper nitrate and silver metal.

$$Cu \quad + \qquad 2\,AgNO_3 \qquad \rightarrow \qquad Cu(NO_3)_2 \qquad + \qquad 2\,Ag$$

Copper   Silver nitrate (Colourless solution)   Copper nitrate (Blue solution)   Silver (White deposit)

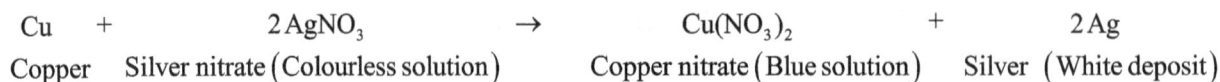

Reaction of Zinc with Copper Sulphate Solution

If a piece of zinc metal is placed in blue coloured solution of copper sulphate for some time, the blue colour of copper sulphate solution fades away. This happens due to the formation of colourless solution of zinc sulphate. You will also observe that during this reaction red brown copper metal deposits on the piece of zinc.

In this reaction zinc is more reactive metal than copper present in copper sulphate solution. So, zinc displaces copper from copper sulphate solution to form zinc sulphate and copper.

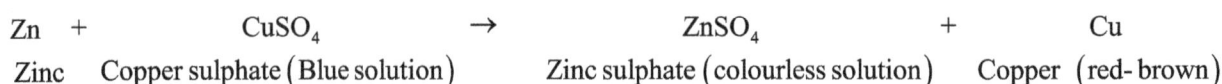

$$Zn \quad + \qquad CuSO_4 \qquad \rightarrow \qquad ZnSO_4 \qquad + \qquad Cu$$

Zinc   Copper sulphate (Blue solution)   Zinc sulphate (colourless solution)   Copper (red-brown)

Reaction of Iron with Copper Sulphate Solution

If iron fillings are placed in the blue coloured solution of $CuSO_4$ for some time, the blue colour of copper sulphate solution turns into greenish colour and red brown precipitate of copper get deposited over iron fillings.

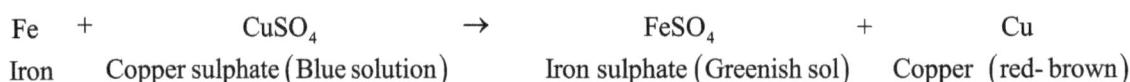

$$Fe \quad + \qquad CuSO_4 \qquad \rightarrow \qquad FeSO_4 \qquad + \qquad Cu$$

Iron   Copper sulphate (Blue solution)   Iron sulphate (Greenish sol)   Copper (red-brown)

Reaction of Metals with Chlorine

All metals react with chlorine to form ionic metal chlorides. For example:

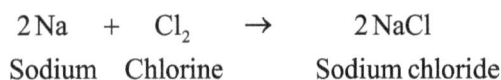

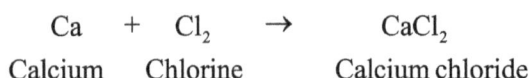

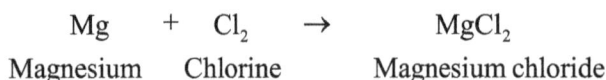

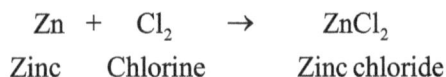

$$2\,Na \quad + \quad Cl_2 \quad \rightarrow \quad 2\,NaCl$$

Sodium   Chlorine   Sodium chloride

$$Ca \quad + \quad Cl_2 \quad \rightarrow \quad CaCl_2$$

Calcium   Chlorine   Calcium chloride

$$Mg \quad + \quad Cl_2 \quad \rightarrow \quad MgCl_2$$

Magnesium   Chlorine   Magnesium chloride

$$Zn \quad + \quad Cl_2 \quad \rightarrow \quad ZnCl_2$$

Zinc   Chlorine   Zinc chloride

Reaction of Metals with Hydrogen

Only a few metals like Na, K, Ca and Mg react with hydrogen to form metal hydrides.

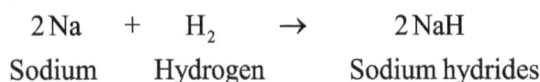

$$2\,Na \quad + \quad H_2 \quad \rightarrow \quad 2\,NaH$$

Sodium   Hydrogen   Sodium hydrides

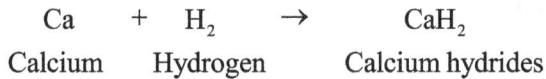

$$Ca \quad + \quad H_2 \quad \rightarrow \quad CaH_2$$

Calcium   Hydrogen   Calcium hydrides

METALLURGY

Metallurgy is the science of extracting metals from their ores and modifying the metals for use. Metallurgy customarily refers to commercial as opposed to laboratory methods. It also concerns the chemical, physical, and atomic properties and structures of metals and the principles whereby metals are combined to form alloys.

The present-day use of metals is the culmination of a long path of development extending over approximately 6,500 years. It is generally agreed that the first known metals were gold, silver, and copper, which occurred in the native or metallic state, of which the earliest were in all probability nuggets of gold found in the sands and gravels of riverbeds. Such native metals became known and were appreciated for their ornamental and utilitarian values during the latter part of the Stone Age.

Earliest Development

Gold can be agglomerated into larger pieces by cold hammering, but native copper cannot, and an essential step toward the Metal Age was the discovery that metals such as copper could be fashioned into shapes by melting and casting in molds; among the earliest known products of this type are copper axes cast in the Balkans in the 4th millennium BCE. Another step was the discovery that metals could be recovered from metal-bearing minerals. These had been collected and could be distinguished on the basis of colour, texture, weight, and flame colour and smell when heated. The notably greater yield obtained by heating native copper with associated oxide minerals may have led to the smelting process, since these oxides are easily reduced to metal in a charcoal bed at temperatures in excess of 700 °C (1,300 °F), as the reductant, carbon monoxide, becomes increasingly stable. In order to effect the agglomeration and separation of melted or smelted copper from its associated minerals, it was necessary to introduce iron oxide as a flux. This further step forward can be attributed to the presence of iron oxide gossan minerals in the weathered upper zones of copper sulfide deposits.

Bronze

In many regions, copper-arsenic alloys, of superior properties to copper in both cast and wrought form, were produced in the next period. This may have been accidental at first, owing to the similarity in colour and flame colour between the bright green copper carbonate mineral malachite and the weathered products of such copper-arsenic sulfide minerals as enargite, and it may have been followed later by the purposeful selection of arsenic compounds based on their garlic odour when heated.

Arsenic contents varied from 1 to 7 percent, with up to 3 percent tin. Essentially arsenic-free copper alloys with higher tin content—in other words, true bronze—seem to have appeared between 3000 and 2500 BCE, beginning in the Tigris-Euphrates delta. The discovery of the value of tin may have occurred through the use of stannite, a mixed sulfide of copper, iron, and tin, although this mineral is not as widely available as the principal tin mineral, cassiterite, which must have been the eventual source of the metal. Cassiterite is strikingly dense and occurs as pebbles in alluvial deposits together with arsenopyrite and gold; it also occurs to a degree in the iron oxide gossans mentioned above.

While there may have been some independent development of bronze in varying localities, it is most likely that the bronze culture spread through trade and the migration of peoples from the Middle East to Egypt, Europe, and possibly China. In many civilizations the production of copper, arsenical copper and tin bronze continued together for some time. The eventual disappearance of copper-arsenic alloys is difficult to explain. Production may have been based on minerals that were not widely available and became scarce, but the relative scarcity of tin minerals did not prevent a substantial trade in that metal over considerable distances. It may be that tin bronzes were eventually preferred owing to the chance of contracting arsenic poisoning from fumes produced by the oxidation of arsenic-containing minerals.

As the weathered copper ores in given localities were worked out, the harder sulfide ores beneath were mined and smelted. The minerals involved, such as chalcopyrite, a copper-iron sulfide, needed an oxidizing roast to remove sulfur as sulfur dioxide and yield copper oxide. This not only required greater metallurgical skill but also oxidized the intimately associated iron, which, combined with the use of iron oxide fluxes and the stronger reducing conditions produced by improved smelting furnaces, led to higher iron contents in the bronze.

Iron

It is not possible to mark a sharp division between the Bronze Age and the Iron Age. Small pieces of iron would have been produced in copper smelting furnaces as iron oxide fluxes and iron-bearing copper sulfide ores were used. In addition, higher furnace temperatures would have created more strongly reducing conditions (that is to say, a higher carbon monoxide content in the furnace gases). An early piece of iron from a trackway in the province of Drenthe, Netherlands, has been dated to 1350 BCE, a date normally taken as the Middle Bronze Age for this area. In Anatolia, on the other hand, iron was in use as early as 2000 BCE. There are also occasional references to iron in even earlier periods, but this material was of meteoric origin.

Once a relationship had been established between the new metal found in copper smelts and the ore added as flux, the operation of furnaces for the production of iron alone naturally followed. Certainly, by 1400 BCE in Anatolia, iron was assuming considerable importance, and by 1200–1000 BCE it was being fashioned on quite a large scale into weapons, initially dagger blades. For this reason, 1200 BCE has been taken as the beginning of the Iron Age. Evidence from excavations indicates that the art of iron making originated in the mountainous country to the south of the Black Sea, an area dominated by the Hittites. Later the art apparently spread to the Philistines, for crude furnaces dating from 1200 BCE have been unearthed at Gerar, together with a number of iron objects.

Smelting of iron oxide with charcoal demanded a high temperature, and, since the melting temperature of iron at 1,540 °C (2,800 °F) was not attainable then, the product was merely a spongy mass of pasty globules of metal intermingled with a semiliquid slag. This product, later known as bloom, was hardly usable as it stood, but repeated reheating and hot hammering eliminated much of the slag, creating wrought iron, a much better product.

The properties of iron are much affected by the presence of small amounts of carbon, with large increases in strength associated with contents of less than 0.5 percent. At the temperatures then attainable—about 1,200 °C (2,200 °F)—reduction by charcoal produced an almost pure iron, which was soft and of limited use for weapons and tools, but when the ratio of fuel to ore was increased and furnace drafting improved with the invention of better bellows, more carbon was absorbed by the iron. This resulted in blooms and iron products with a range of carbon contents, making it difficult to determine the period in which iron may have been purposely strengthened by carburizing, or reheating the metal in contact with excess charcoal.

Carbon-containing iron had the further great advantage that, unlike bronze and carbon-free iron, it could be made still harder by quenching—i.e., rapid cooling by immersion in water. There is no evidence for the use of this hardening process during the early Iron Age, so that it must have been either unknown then or not considered advantageous, in that quenching renders iron very brittle and has to be followed by tempering, or reheating at a lower temperature, to restore toughness. What seems to have been established early on was a practice of repeated cold forging and annealing at 600–700 °C (1,100–1,300 °F), a temperature naturally achieved in a simple fire. This practice is common in parts of Africa even today.

By 1000 BCE iron was beginning to be known in central Europe. Its use spread slowly westward. Iron making was fairly widespread in Great Britain at the time of the Roman invasion in 55 BCE. In Asia iron was also known in ancient times, in China by about 700 BCE.

Brass

While some zinc appears in bronzes dating from the Bronze Age, this was almost certainly an accidental inclusion, although it may foreshadow the complex ternary alloys of the early Iron Age, in which substantial amounts of zinc as well as tin may be found. Brass, as an alloy of copper and zinc without tin, did not appear in Egypt until about 30 BCE, but after this it was rapidly adopted throughout the Roman world, for example, for currency. It was made by the calamine process, in which zinc carbonate or zinc oxide were added to copper and melted under a charcoal cover in order to produce reducing conditions. The general establishment of a brass industry was one of the important metallurgical contributions made by the Romans.

Precious Metals

Bronze, iron, and brass were, then, the metallic materials on which successive peoples built their civilizations and of which they made their implements for both war and peace. In addition, by 500 BCE, rich lead-bearing silver mines had opened in Greece. Reaching depths of several hundred metres, these mines were vented by drafts provided by fires lit at the bottom of the shafts. Ores were hand-sorted, crushed, and washed with streams of water to separate valuable minerals from the barren, lighter materials. Because these minerals were principally sulfides, they were roasted to form oxides and were then smelted to recover a lead-silver alloy.

Lead was removed from the silver by cupellation, a process of great antiquity in which the alloy was melted in a shallow porous clay or bone-ash receptacle called a cupel. A stream of air over the molten mass preferentially oxidized the lead. Its oxide was removed partially by skimming the molten surface; the remainder was absorbed into the porous cupel. Silver metal and any gold were retained on the cupel. The lead from the skimmings and discarded cupels was recovered as metal upon heating with charcoal.

Native gold itself often contained quite considerable quantities of silver. These silver-gold alloys known as electrum, may be separated in a number of ways, but presumably the earliest was by heating in a crucible with common salt. In time and with repetitive treatments, the silver was converted into silver chloride, which passed into the molten slag, leaving a purified gold. Cupellation was also employed to remove from the gold such contaminates as copper, tin, and lead. Gold, silver, and lead were used for artistic and religious purposes, personal adornment, household utensils, and equipment for the chase.

In the thousand years between 500 BCE and 500 CE, a vast number of discoveries of significance to the growth of metallurgy were made. The Greek mathematician and inventor Archimedes, for example, demonstrated that the purity of gold could be measured by determining its weight and the quantity of water displaced upon immersion—that is, by determining its density. In the pre-Christian portion of the period, the first important steel production was started in India, using a process already known to ancient Egyptians. Wootz steel, as it was called, was prepared as sponge (porous) iron in a unit not unlike a bloomery. The product was hammered while hot to expel slag, broken up, then sealed with wood chips in clay containers and heated until the pieces of iron absorbed carbon and melted, converting it to steel of homogeneous composition containing 1 to 1.6 percent carbon. The steel pieces could then be heated and forged to bars for later use in fashioning articles, such as the famous Damascus swords made by medieval Arab armourers.

Arsenic, zinc, antimony, and nickel may well have been known from an early date but only in the alloy state. By 100 BCE mercury was known and was produced by heating the sulfide mineral cinnabar and condensing the vapours. Its property of amalgamating (mixing or alloying) with various metals was employed for their recovery and refining. Lead was beaten into sheets and pipes, the pipes being used in early water systems. The metal tin was available and Romans had learned to use it to line food containers. Although the Romans made no extraordinary metallurgical discoveries, they were responsible for, in addition to the establishment of the brass industry, contributing toward improved organization and efficient administration in mining.

Beginning about the 6th century, and for the next thousand years, the most meaningful developments in metallurgy centred on iron making. Great Britain, where iron ore was plentiful, was an important iron-making region. Iron weapons, agricultural implements, domestic articles, and even personal adornments were made. Fine-quality cutlery was made near Sheffield. Monasteries were often centres of learning of the arts of metalworking. Monks became well known for their iron making and bell founding, the products made either being utilized in the monasteries, disposed of locally, or sold to merchants for shipment to more distant markets. In 1408 the bishop of Durham established the first water-powered bloomery in Britain, with the power apparently operating the bellows. Once power of this sort became available, it could be applied to a range of operations and enable the hammering of larger blooms.

In Spain, another iron-making region, the Catalan forge had been invented, and its use later spread to other areas. A hearth type of furnace, it was built of stone and was charged with iron ore, flux, and charcoal. The charcoal was kept ignited with air from a bellows blown through a bottom nozzle, or tuyere. The bloom that slowly collected at the bottom was removed and upon frequent reheating and forging was hammered into useful shapes. By the 14th century the furnace was greatly enlarged in height and capacity.

If the fuel-to-ore ratio in such a furnace was kept high, and if the furnace reached temperatures sufficiently hot for substantial amounts of carbon to be absorbed into the iron, then the melting point of the metal would be lowered and the bloom would melt. This would dissolve even more carbon, producing a liquid cast iron of up to 4 percent carbon and with a relatively low melting temperature of 1,150 °C (2,100 °F). The cast iron would collect in the base of the furnace, which technically would be a blast furnace rather than a bloomery in that the iron would be withdrawn as a liquid rather than a solid lump.

While the Iron Age peoples of Anatolia and Europe on occasion may have accidently made cast iron, which is chemically the same as blast-furnace iron, the Chinese were the first to realize its advantages. Although brittle and lacking the strength, toughness, and workability of steel, it was useful for making cast bowls and other vessels. In fact the Chinese, whose Iron Age began about 500 BCE, appear to have learned to oxidize the carbon from cast iron in order to produce steel or wrought iron indirectly, rather than through the direct method of starting from low-carbon iron.

During the 16th century, metallurgical knowledge was recorded and made available. Two books were especially influential. One, by the Italian Vannoccio Biringuccio, was entitled De la pirotechnia. The other, by the German Georgius Agricola, was entitled De re metallica. Biringuccio was essentially a metalworker, and his book dealt with smelting, refining, and assay methods (methods for determining the metal content of ores) and covered metal casting, molding, core making, and the production of such commodities as cannons and cast-iron cannonballs. His was the first methodical description of foundry practice.

Agricola, on the other hand, was a miner and an extractive metallurgist; his book considered prospecting and surveying in addition to smelting, refining, and assay methods. He also described the processes used for crushing and concentrating the ore and then, in some detail, the methods of assaying to determine whether ores were worth mining and extracting. Some of the metallurgical practices he described are retained in principle today.

Ferrous Metals

From 1500 to the 20th century, metallurgical development was still largely concerned with improved technology in the manufacture of iron and steel. In England, the gradual exhaustion of timber led first to prohibitions on cutting of wood for charcoal and eventually to the introduction of coke, derived from coal, as a more efficient fuel. Thereafter, the iron industry expanded rapidly in Great Britain, which became the greatest iron producer in the world. The crucible process for making steel, introduced in England in 1740, by which bar iron and added materials were placed in clay crucibles heated by coke fires, resulted in the first reliable steel made by a melting process.

One difficulty with the bloomery process for the production of soft bar iron was that, unless the temperature was kept low (and the output therefore small), it was difficult to keep the carbon content low enough so that the metal remained ductile. This difficulty was overcome by melting high-carbon pig iron from the blast furnace in the puddling process, invented in Great Britain in 1784. In it, melting was accomplished by drawing hot gases over a charge of pig iron and iron ore held on the furnace hearth. During its manufacture the product was stirred with iron rabbles (rakes), and as it became pasty with loss of carbon, it was worked into balls, which were subsequently forged or rolled to a useful shape. The product which came to be known as wrought iron, was low in elements that contributed to the brittleness of pig iron and contained enmeshed slag particles that became elongated fibres when the metal was forged. Later, the use of a rolling mill equipped with grooved rolls to make wrought-iron bars was introduced.

The most important development of the 19th century was the large-scale production of cheap steel. Prior to about 1850, the production of wrought iron by puddling and of steel by crucible melting had been conducted in small-scale units without significant mechanization. The first change was the development of the open-hearth furnace by William and Friedrich Siemens in Britain and by Pierre and Émile Martin in France. Employing the regenerative principle, in which outgoing combusted gases are used to heat the next cycle of fuel gas and air, this enabled high temperatures to be achieved while saving on fuel. Pig iron could then be taken through to molten iron or low-carbon steel without solidification, scrap could be added and melted, and iron ore could be melted into the slag above the metal to give a relatively rapid oxidation of carbon and silicon—all on a much enlarged scale. Another major advance was Henry Bessemer's process, patented in 1855 and first operated in 1856, in which air was blown through molten pig iron from tuyeres set into the bottom of a pear-shaped vessel called a converter. Heat released by the oxidation of dissolved silicon, manganese, and carbon was enough to raise the temperature above the melting point of the refined metal (which rose as the carbon content was lowered) and thereby maintain it in the liquid state. Very soon Bessemer had tilting converters producing 5 tons in a heat of one hour, compared with four to six hours for 50 kilograms (110 pounds) of crucible steel and two hours for 250 kilograms of puddled iron.

Neither the open-hearth furnace nor the Bessemer converter could remove phosphorus from the metal, so that low-phosphorus raw materials had to be used. This restricted their use from areas where phosphoric ores, such as those of the Minette range in Lorraine, were a main European source of iron. The problem was solved by Sidney Gilchrist Thomas, who demonstrated in 1876 that a basic furnace lining consisting of calcined dolomite, instead of an acidic lining of siliceous materials, made it possible to use a high-lime slag to dissolve the phosphates formed by the oxidation of phosphorus in the pig iron. This principle was eventually applied to both open-hearth furnaces and Bessemer converters.

As steel was now available at a fraction of its former cost, it saw an enormously increased use for engineering and construction. Soon after the end of the century it replaced wrought iron in virtually every field. Then, with the availability of electric power, electric-arc furnaces were introduced for making special and high-alloy steels. The next significant stage was the introduction of cheap oxygen, made possible by the invention of the Linde-Frankel cycle for the liquefaction and fractional distillation of air. The Linz-Donawitz process, invented in Austria shortly after World War II, used oxygen supplied as a gas from a tonnage oxygen plant, blowing it at supersonic velocity into the top of the molten iron in a converter vessel. As the ultimate development of the Bessemer/Thomas process, oxygen blowing became universally employed in bulk steel production.

Light Metals

Another important development of the late 19th century was the separation from their ores, on a substantial scale, of aluminum and magnesium. In the earlier part of the century, several scientists had made small quantities of these light metals, but the most successful was Henri-Éti-enne Sainte-Claire Deville, who by 1855 had developed a method by which cryolite, a double flu-oride of aluminum and sodium, was reduced by sodium metal to aluminum and sodium fluoride. The process was very expensive, but cost was greatly reduced when the American chemist Ham-ilton Young Castner developed an electrolytic cell for producing cheaper sodium in 1886. At the same time, however, Charles M. Hall in the United States and Paul-Louis-Toussaint Héroult in France announced their essentially identical processes for aluminum extraction, which were also based on electrolysis. Use of the Hall-Héroult process on an industrial scale depended on the replacement of storage batteries by rotary power generators; it remains essentially unchanged to this day.

Welding

One of the most significant changes in the technology of metals fabrication has been the introduc-tion of fusion welding during the 20th century. Before this, the main joining processes were rivet-ing and forge welding. Both had limitations of scale, although they could be used to erect substan-tial structures. In 1895 Henry-Louis Le Chatelier stated that the temperature in an oxyacetylene flame was 3,500 °C (6,300 °F), some 1,000 °C higher than the oxyhydrogen flame already in use on a small scale for brazing and welding. The first practical oxyacetylene torch, drawing acetylene from cylinders containing acetylene dissolved in acetone, was produced in 1901. With the avail-ability of oxygen at even lower cost, oxygen cutting and oxyacetylene welding became established procedures for the fabrication of structural steel components.

The metal in a join can also be melted by an electric arc, and a process using a carbon as a negative electrode and the workpiece as a positive first became of commercial interest about 1902. Striking an arc from a coated metal electrode, which melts into the join, was introduced in 1910. Although it was not widely used until some 20 years later, in its various forms it is now responsible for the bulk of fusion welds.

Metallography

The 20th century has seen metallurgy change progressively, from an art or craft to a scientific dis-cipline and then to part of the wider discipline of materials science. In extractive metallurgy, there has been the application of chemical thermodynamics, kinetics, and chemical engineering, which has enabled a better understanding, control, and improvement of existing processes and the gener-ation of new ones. In physical metallurgy, the study of relationships between macrostructure, mi-crostructure, and atomic structure on the one hand and physical and mechanical properties on the other has broadened from metals to other materials such as ceramics, polymers, and composites.

This greater scientific understanding has come largely from a continuous improvement in micro-scopic techniques for metallography, the examination of metal structure. The first true metallog-rapher was Henry Clifton Sorby of Sheffield, England, who in the 1860s applied light microscopy to the polished surfaces of materials such as rocks and meteorites. Sorby eventually succeeded in making photomicrographic records, and by 1885 the value of metallography was appreciated throughout Europe, with particular attention being paid to the structure of steel. For example, there was eventual acceptance, based on micrographic evidence and confirmed by the introduction of X-ray diffraction by William Henry and William Lawrence Bragg in 1913, of the allotropy of iron and its relationship to the hardening of steel. During subsequent years there were advances in the atomic theory of solids; this led to the concept that, in nonplastic materials such as glass, fracture takes place by the propagation of preexisting cracklike defects and that, in metals, deformation takes place by the movement of dislocations, or defects in the atomic arrangement, through the crystalline matrix. Proof of these concepts came with the invention and development of the electron microscope; even more powerful field ion microscopes and high-resolution electron microscopes now make it possible to detect the position of individual atoms.

Another example of the development of physical metallurgy is a discovery that revolutionized the use of aluminum in the 20th century. Originally, most aluminum was used in cast alloys, but the discovery of age hardening by Alfred Wilm in Berlin about 1906 yielded a material that was twice as strong with only a small change in weight. In Wilm's process, a solute such as magnesium or copper is trapped in supersaturated solid solution, without being allowed to precipitate out, by quenching the aluminum from a higher temperature rather than slowly cooling it. The relatively soft aluminum alloy that results can be mechanically formed, but, when left at room temperature or heated at low temperatures, it hardens and strengthens. With copper as the solute, this type of material came to be known by the trade name Duralumin. The advances in metallography described above eventually provided the understanding that age hardening is caused by the dispersion of very fine precipitates from the supersaturated solid solution; this restricts the movement of the dislocations that are essential to crystal deformation and thus raises the strength of the metal. The principles of precipitation hardening have been applied to the strengthening of a large number of alloys.

FERROUS METALLURGY

Bloomery smelting during the Middle Ages.

Ferrous metallurgy is the metallurgy of iron and its alloys. It began far back in prehistory. The earliest surviving iron artifacts, from the 4th millennium BC in Egypt, were made from meteoritic iron-nickel. It is not known when or where the smelting of iron from ores began, but by the end of the 2nd millennium BC iron was being produced from iron ores from Sub-Saharan Africa to China. The use of wrought iron (worked iron) was known by the 1st millennium BC, and its spread marked the Iron Age. During the medieval period, means were found in Europe of producing wrought iron from cast iron (in this context known as pig iron) using finery forges. For all these processes, charcoal was required as fuel.

Steel (with a carbon content between pig iron and wrought iron) was first produced in antiquity as an alloy. Its process of production, Wootz steel, was exported before the 4th century BC from India to ancient China, Africa, the Middle East and Europe. Archaeological evidence of cast iron appears in 5th-century BC China. New methods of producing it by carburizing bars of iron in the cementation process were devised in the 17th century. During the Industrial Revolution, new methods of producing bar iron by substituting coke for charcoal were devised and these were later applied to produce steel, creating a new era of greatly increased use of iron and steel that some contemporaries described as a new Iron Age. In the late 1850s, Henry Bessemer invented a new steelmaking process, that involved blowing air through molten pig iron to burn off carbon, and so to produce mild steel. This and other 19th-century and later steel making processes have displaced wrought iron. Today, wrought iron is no longer produced on a commercial scale having been displaced by the functionally equivalent mild or low carbon steel.

The largest and most modern underground iron ore mine in the world is located in Kiruna, Norrbotten County, Lapland. The mine which is owned by Luossavaara-Kiirunavaara AB, a large Swedish mining company, has an annual production capacity of over 26 million tonnes of iron ore.

Meteoritic Iron

Willamette Meteorite, the sixth largest in the world, is an iron-nickel meteorite.

Iron was extracted from iron–nickel alloys, which comprise about 6% of all meteorites that fall on the Earth. That source can often be identified with certainty because of the unique crystalline features (Widmanstätten patterns) of that material, which are preserved when the metal is worked cold or at low temperature. Those artifacts include, for example, a bead from the 5th millennium BC found in Iran and spear tips and ornaments from Ancient Egypt and Sumer around 4000 BC.

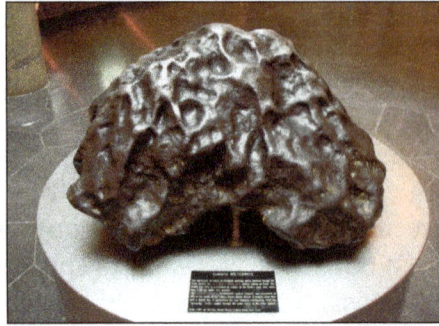

Iron meteorites consist overwhelmingly of nickel-iron alloys. The metal taken from these meteorites is known as meteoritic iron and was one of the earliest sources of usable iron available to humans.

These early uses appear to have been largely ceremonial or ornamental. Meteoritic iron is very rare, and the metal was probably very expensive, perhaps more expensive than gold. The early Hittites are known to have bartered iron (meteoritic or smelted) for silver, at a rate of 40 times the iron's weight, with the Old Assyrian Empire in the first centuries of the second millennium BC.

Meteoric iron was also fashioned into tools in the Arctic, about the year 1000, when the Thule people of Greenland began making harpoons, knives, ulus and other edged tools from pieces of the Cape York meteorite. Typically pea-size bits of metal were cold-hammered into disks and fitted to a bone handle. These artifacts were also used as trade goods with other Arctic peoples: tools made from the Cape York meteorite have been found in archaeological sites more than 1,000 miles (1,600 km) distant. When the American polar explorer Robert Peary shipped the largest piece of the meteorite to the American Museum of Natural History in New York City in 1897, it still weighed over 33 tons. Another example of a late use of meteoritic iron is an adze from around 1000 AD found in Sweden.

Native Iron

Native iron in the metallic state occurs rarely as small inclusions in certain basalt rocks. Besides meteoritic iron, Thule people of Greenland have used native iron from the Disko region.

Iron Smelting and the Iron Age

Iron smelting—the extraction of usable metal from oxidized iron ores—is more difficult than tin and copper smelting. While these metals and their alloys can be cold-worked or melted in relatively simple furnaces (such as the kilns used for pottery) and cast into molds, smelted iron requires hot-working and can be melted only in specially designed furnaces. Iron is a common impurity in copper ores and iron ore was sometimes used as a flux, thus it is not surprising that humans mastered the technology of smelted iron only after several millennia of bronze metallurgy.

The place and time for the discovery of iron smelting is not known, partly because of the difficulty of distinguishing metal extracted from nickel-containing ores from hot-worked meteoritic iron. The archaeological evidence seems to point to the Middle East area, during the Bronze Age in the 3rd millennium BC. However, wrought iron artifacts remained a rarity until the 12th century BC.

The Iron Age is conventionally defined by the widespread replacement of bronze weapons and tools with those of iron and steel. That transition happened at different times in different places, as the technology spread. Mesopotamia was fully into the Iron Age by 900 BC. Although Egypt produced iron artifacts, bronze remained dominant until its conquest by Assyria in 663 BC. The Iron Age began in India about 1200 BC, in Central Europe about 600 BC, and in China about 300 BC. Around 500 BC, the Nubians who had learned from the Assyrians the use of iron and were expelled from Egypt, became major manufacturers and exporters of iron.

Ancient Near East

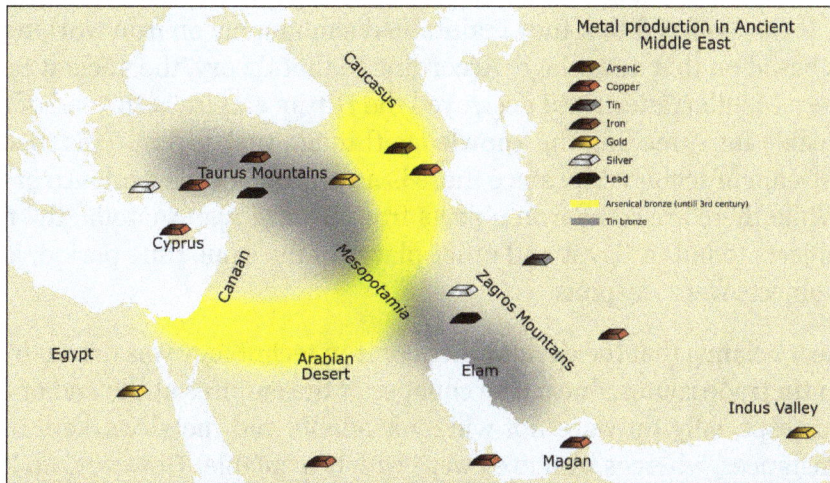

Mining areas of the ancient Middle East. Boxes colors: arsenic is in brown, copper in red, tin in grey, iron in reddish brown, gold in yellow, silver in white and lead in black. Yellow area stands for arsenic bronze, while grey area stands for tin bronze.

One of the earliest smelted iron artifacts, a dagger with an iron blade found in a Hattic tomb in Anatolia, dated from 2500 BC. About 1500 BC, increasing numbers of non-meteoritic, smelted iron objects appeared in Mesopotamia, Anatolia and Egypt. Nineteen meteoric iron objects were found in the tomb of Egyptian ruler Tutankhamun, who died in 1323 BC, including an iron dagger with a golden hilt, an Eye of Horus, the mummy's head-stand and sixteen models of an artisan's tools. An Ancient Egyptian sword bearing the name of pharaoh Merneptah as well as a battle axe with an iron blade and gold-decorated bronze shaft were both found in the excavation of Ugarit.

Although iron objects dating from the Bronze Age have been found across the Eastern Mediterranean, bronzework appears to have greatly predominated during this period. By the 12th century BC, iron smelting and forging, of weapons and tools, was common from Sub-Saharan Africa through India. As the technology spread, iron came to replace bronze as the dominant metal used for tools and weapons across the Eastern Mediterranean (the Levant, Cyprus, Greece, Crete, Anatolia and Egypt).

Iron was originally smelted in bloomeries, furnaces where bellows were used to force air through a pile of iron ore and burning charcoal. The carbon monoxide produced by the charcoal reduced the iron oxide from the ore to metallic iron. The bloomery, however, was not hot enough to melt the iron, so the metal collected in the bottom of the furnace as a spongy mass, or *bloom*. Workers then repeatedly beat and folded it to force out the molten slag. This laborious, time-consuming process produced wrought iron, a malleable but fairly soft alloy.

Concurrent with the transition from bronze to iron was the discovery of carburization, the process of adding carbon to wrought iron. While the iron bloom contained some carbon, the subsequent hot-working oxidized most of it. Smiths in the Middle East discovered that wrought iron could be turned into a much harder product by heating the finished piece in a bed of charcoal, and then quenching it in water or oil. This procedure turned the outer layers of the piece into steel, an alloy of iron and iron carbides, with an inner core of less brittle iron.

Theories on the Origin of Iron Smelting

The development of iron smelting was traditionally attributed to the Hittites of Anatolia of the Late Bronze Age. It was believed that they maintained a monopoly on iron working, and that their empire had been based on that advantage. According to that theory, the ancient Sea Peoples, who invaded the Eastern Mediterranean and destroyed the Hittite empire at the end of the Late Bronze Age, were responsible for spreading the knowledge through that region. This theory is no longer held in the mainstream of scholarship, since there is no archaeological evidence of the alleged Hittite monopoly. While there are some iron objects from Bronze Age Anatolia, the number is comparable to iron objects found in Egypt and other places of the same time period, and only a small number of those objects were weapons.

A more recent theory claims that the development of iron technology was driven by the disruption of the copper and tin trade routes, due to the collapse of the empires at the end of the Late Bronze Age. These metals especially tin were not widely available and metal workers had to transport them over long distances, whereas iron ores were widely available. However, no known archaeological evidence suggests a shortage of bronze or tin in the Early Iron Age. Bronze objects remained abundant, and these objects have the same percentage of tin as those from the Late Bronze Age.

Indian Sub-continent

The history of metallurgy in the Indian subcontinent began in the 2nd millennium BC. Archaeological sites in Gangetic plains have yielded iron implements dated between 1800–1200 BC. By the early 13th century BC, iron smelting was practiced on a large scale in India. In Southern India (present day Mysore) iron was in use 12th to 11th centuries BC. The technology of iron metallurgy advanced in the politically stable Maurya period and during a period of peaceful settlements in the 1st millennium BC.

The Iron pillar of Delhi.

Iron artifacts such as spikes, knives, daggers, arrow-heads, bowls, spoons, saucepans, axes, chisels, tongs, door fittings, etc., dated from 600 to 200 BC, have been discovered at several archaeological sites of India. The Greek historian Herodotus wrote the first western account of the use of iron in India. The Indian mythological texts, the Upanishads, have mentions of weaving, pottery and metallurgy, as well. The Romans had high regard for the excellence of steel from India in the time of the Gupta Empire.

Dagger and its scabbard, India, 17th–18th century. Blade: Damascus steel inlaid with gold; hilt: jade; scabbard: steel with engraved, chased and gilded decoration.

Perhaps as early as 500 BC, although certainly by 200 AD, high-quality steel was produced in southern India by the crucible technique. In this system, high-purity wrought iron, charcoal, and glass were mixed in a crucible and heated until the iron melted and absorbed the carbon. Iron chain was used in Indian suspension bridges as early as the 4th century.

Wootz steel was produced in India and Sri Lanka from around 300 BC. Wootz steel is famous from Classical Antiquity for its durability and ability to hold an edge. When asked by King Porus to select a gift, Alexander is said to have chosen, over gold or silver, thirty pounds of steel. Wootz steel was originally a complex alloy with iron as its main component together with various trace elements. Recent studies have suggested that its qualities may have been due to the formation of carbon nanotubes in the metal. According to Will Durant, the technology passed to the Persians and from them to Arabs who spread it through the Middle East. In the 16th century, the Dutch carried the technology from South India to Europe, where it was mass-produced.

Steel was produced in Sri Lanka from 300 BC by furnaces blown by the monsoon winds. The furnaces were dug into the crests of hills, and the wind was diverted into the air vents by long trenches. This arrangement created a zone of high pressure at the entrance, and a zone of low pressure at the top of the furnace. The flow is believed to have allowed higher temperatures than bellows-driven furnaces could produce, resulting in better-quality iron. Steel made in Sri Lanka was traded extensively within the region and in the Islamic world.

One of the world's foremost metallurgical curiosities is an iron pillar located in the Qutb complex in Delhi. The pillar is made of wrought iron (98% Fe), is almost seven meters high and weighs more than six tonnes. The pillar was erected by Chandragupta II Vikramaditya and has withstood 1,600 years of exposure to heavy rains with relatively little corrosion.

China

Historians debate whether bloomery-based ironworking ever spread to China from the Middle East. One theory suggests that metallurgy was introduced through Central Asia. In 2008, two iron fragments were excavated at the Mogou site, in Gansu. They have been dated to the 14th century BC, belonging to the period of Siwa culture. One of the fragments was made of bloomery iron rather than meteoritic iron.

The earliest cast iron artifacts, dating to 5th century BC, were discovered by archaeologists in what is now modern Luhe County, Jiangsu in China. Cast iron was used in ancient China for warfare, agriculture and architecture. Around 500 BC, metalworkers in the southern state of Wu achieved a temperature of 1130 °C. At this temperature, iron combines with 4.3% carbon and melts. The liquid iron can be cast into molds, a method far less laborious than individually forging each piece of iron from a bloom.

Cast iron is rather brittle and unsuitable for striking implements. It can, however, be *decarburized* to steel or wrought iron by heating it in air for several days. In China, these iron working methods spread northward, and by 300 BC, iron was the material of choice throughout China for most tools and weapons. A mass grave in Hebei province, dated to the early 3rd century BC, contains several soldiers buried with their weapons and other equipment. The artifacts recovered from this grave are variously made of wrought iron, cast iron, malleabilized cast iron, and quench-hardened steel, with only a few, probably ornamental bronze weapons.

An illustration of furnace bellows operated by waterwheels, from the *Nong Shu*, by Wang Zhen, 1313 AD, during the Yuan Dynasty in China.

During the Han Dynasty (202 BC–220 AD), the government established ironworking as a state monopoly (repealed during the latter half of the dynasty and returned to private entrepreneurship) and built a series of large blast furnaces in Henan province, each capable of producing several tons of iron per day. By this time, Chinese metallurgists had discovered how to fine molten pig iron, stirring it in the open air until it lost its carbon and could be hammered (wrought). (In modern Mandarin-Chinese, this process is now called *chao*, literally, stir frying.) By the 1st century BC, Chinese metallurgists had found that wrought iron and cast iron could be melted together to yield an alloy of intermediate carbon content, that is steel. According to legend, the sword of Liu Bang, the first Han emperor, was made in this fashion. Some texts of the era mention "harmonizing the hard and the soft" in the context of ironworking; the phrase may refer to this process. The ancient city of Wan (Nanyang) from the Han period forward was a major center of the iron and steel industry. Along with their original methods of forging steel, the Chinese had also adopted the production methods of creating Wootz steel, an idea imported from India to China by the 5th century AD. During the Han Dynasty, the Chinese were also the first to apply hydraulic power (i.e. a waterwheel) in working the bellows of the blast furnace. This was recorded in the year 31 AD, as an innovation by the Chinese mechanical engineer and politician Du Shi, Prefect of Nanyang. Although Du Shi was the first to apply water power to bellows in metallurgy, the first drawn and printed illustration of its operation with water power appeared in 1313 AD, in the Yuan Dynasty era text called the *Nong Shu*. In the 11th century, there is evidence of the production of steel in Song China using two techniques: a "berganesque" method that produced inferior, heterogeneous steel and a precursor to the modern Bessemer process that utilized partial decarbonization via repeated forging under a cold blast. By the 11th century, there was a large amount of deforestation in China due to the iron industry's demands for charcoal. By this time however, the Chinese had learned to use bituminous coke to replace charcoal, and with this switch in resources many acres of prime timberland in China were spared.

Iron Age Europe

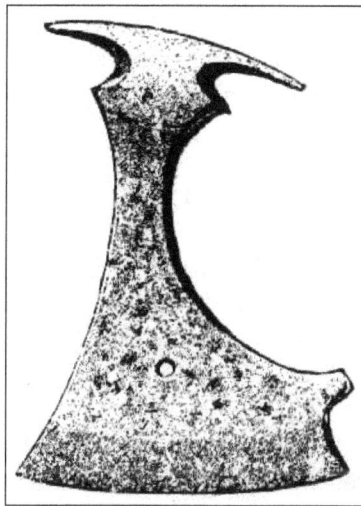

Axe made of iron, dating from
Swedish Iron Age, found at Gotland, Sweden.

Iron working was introduced to Greece in the late 10th century BC. The earliest marks of Iron Age in Central Europe are artifacts from the Hallstatt C culture (8th century BC). Throughout the 7th to 6th centuries BC, iron artifacts remained luxury items reserved for an elite. This changed dramatically shortly after 500 BC with the rise of the La Tène culture, from which time iron metallurgy also became common in Northern Europe and Britain. The spread of ironworking in Central and Western Europe is associated with Celtic expansion. By the 1st century BC, Noric steel was famous for its quality and sought-after by the Roman military.

Africa South of the Sahara

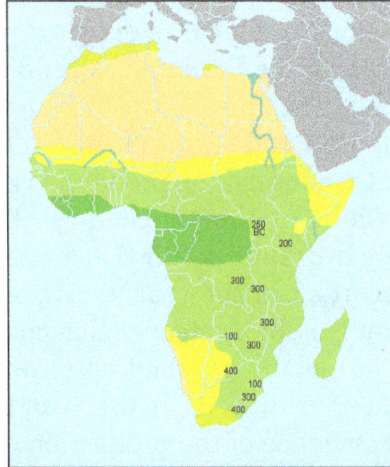

Iron Age finds in East and Southern Africa, corresponding to the early 1st millennium AD Bantu expansion.

Inhabitants at Termit, in eastern Niger became the first iron smelting people in West Africa around 1500 BC. Iron and copper working spread southward through the continent, reaching the Cape around AD 200. The widespread use of iron revolutionized the Bantu-speaking farming communities who adopted it, driving out and absorbing the rock tool using hunter-gatherer societies they encountered as they expanded to farm wider areas of savanna. The technologically superior Bantu-speakers spread across southern Africa and became wealthy and powerful, producing iron for tools and weapons in large, industrial quantities.

In the region of the Air Mountains in Niger there are signs of independent copper smelting between 2500–1500 BC. The process was not in a developed state, indicating smelting was not foreign. It became mature about 1500 BC.

Similarly, smelting in bloomery-type furnaces in West Africa and forging for tools appear in the Nok culture in Africa by 500 BC. The earliest records of bloomery-type furnaces in East Africa are discoveries of smelted iron and carbon in Nubia and Axum that date back between 1000–500 BC. Particularly in Meroe, there are known to have been ancient bloomeries that produced metal tools for the Nubians and Kushites and produced surplus for their economy.

Medieval Islamic World

Iron technology was further advanced by several inventions in medieval Islam, during the Islamic Golden Age. These included a variety of water-powered and wind-powered industrial mills for metal production, including geared gristmills and forges. By the 11th century, every province throughout the Muslim world had these industrial mills in operation, from Islamic Spain and North Africa in the west to the Middle East and Central Asia in the east. There are also 10th-century references to cast iron, as well as archeological evidence of blast furnaces being used in the Ayyubid and Mamluk empires from the 11th century, thus suggesting a diffusion of Chinese metal technology to the Islamic world.

Geared gristmills were invented by Muslim engineers, and were used for crushing metallic ores before extraction. Gristmills in the Islamic world were often made from both watermills and windmills. In order to adapt water wheels for gristmilling purposes, cams were used for raising and releasing trip hammers. The first forge driven by a hydropowered water mill rather than manual labour was invented in the 12th century Islamic Spain.

One of the most famous steels produced in the medieval Near East was Damascus steel used for swordmaking, and mostly produced in Damascus, Syria, in the period from 900 to 1750. This was produced using the crucible steel method, based on the earlier Indian wootz steel. This process was adopted in the Middle East using locally produced steels. The exact process remains unknown, but it allowed carbides to precipitate out as micro particles arranged in sheets or bands within the body of a blade. Carbides are far harder than the surrounding low carbon steel, so swordsmiths could produce an edge that cut hard materials with the precipitated carbides, while the bands of softer steel let the sword as a whole remain tough and flexible. A team of researchers based at the Technical University of Dresden that uses X-rays and electron microscopy to examine Damascus steel discovered the presence of cementite nanowires and carbon nanotubes. Peter Paufler, a member of the Dresden team, says that these nanostructures give Damascus steel its distinctive properties and are a result of the forging process.

Medieval and Early Modern Europe

There was no fundamental change in the technology of iron production in Europe for many centuries. European metal workers continued to produce iron in bloomeries. However, the Medieval period brought two developments—the use of water power in the bloomery process in various places, and the first European production in cast iron.

Powered Bloomeries

Sometime in the medieval period, water power was applied to the bloomery process. It is possible that this was at the Cistercian Abbey of Clairvaux as early as 1135, but it was certainly in use in early 13th century France and Sweden. In England, the first clear documentary evidence for this is the accounts of a forge of the Bishop of Durham, near Bedburn in 1408, but that was certainly not the first such ironworks. In the Furness district of England, powered bloomeries were in use into the beginning of the 18th century, and near Garstang until about 1770.

The Catalan Forge was a variety of powered bloomery. Bloomeries with hot blast were used in upstate New York in the mid-19th century.

Blast Furnace

The preferred method of iron production in Europe until the development of the puddling process in 1783-84. Cast iron development lagged in Europe because wrought iron was the desired product and the intermediate step of producing cast iron involved an expensive blast furnace and further refining of pig iron to cast iron, which then required a labor and capital intensive conversion to wrought iron.

Through a good portion of the Middle Ages, in Western Europe, iron was still being made by the working of iron blooms into wrought iron. Some of the earliest casting of iron in Europe occurred in Sweden, in two sites, Lapphyttan and Vinarhyttan, between 1150 and 1350. Some scholars have speculated the practice followed the Mongols across Russia to these sites, but there is no clear proof of this hypothesis, and it would certainly not explain the pre-Mongol datings of many of these iron-production centres. In any event, by the late 14th century, a market for cast iron goods began to form, as a demand developed for cast iron cannonballs.

Finery Forge

An alternative method of decarburising pig iron was the finery forge, which seems to have been devised in the region around Namur in the 15th century. By the end of that century, this Walloon process spread to the *Pay de Bray* on the eastern boundary of Normandy, and then to England, where it became the main method of making wrought iron by 1600. It was introduced to Sweden by Louis de Geer in the early 17th century and was used to make the oregrounds iron favoured by English steelmakers.

A variation on this was the German forge. This became the main method of producing bar iron in Sweden.

Cementation Process

In the early 17th century, ironworkers in Western Europe had developed the cementation process for carburizing wrought iron. Wrought iron bars and charcoal were packed into stone boxes, then sealed with clay to be held at a red heat continually tended in an oxygen-free state immersed in nearly pure carbon (charcoal) for up to a week. During this time, carbon diffused into the surface layers of the iron, producing *cement steel* or *blister steel*—also known as case hardened, where the portions wrapped in iron (the pick or axe blade) became harder, than say an axe hammer-head or shaft socket which might be insulated by clay to keep them from the carbon source. The earliest place where this process was used in England was at Coalbrookdale from 1619, where Sir Basil Brooke had two cementation furnaces. For a time in the 1610s, he owned a patent on the process, but had to surrender this in 1619. He probably used Forest of Dean iron as his raw material, but it was soon found that oregrounds iron was more suitable. The quality of the steel could be improved by faggoting, producing the so-called shear steel.

Crucible Steel

In the 1740s, Benjamin Huntsman found a means of melting blister steel, made by the cementation process, in crucibles. The resulting crucible steel, usually cast in ingots, was more homogeneous than blister steel.

Transition to Coke in England

Early iron smelting used charcoal as both the heat source and the reducing agent. By the 18th century, the availability of wood for making charcoal was limiting the expansion of iron production, so that England became increasingly dependent for a considerable part of the iron required by its industry, on Sweden (from the mid-17th century) and then from about 1725 also on Russia.

Smelting with coal (or its derivative coke) was a long sought objective. The production of pig iron with coke was probably achieved by Dud Dudley in the 1620s, and with a mixed fuel made from coal and wood again in the 1670s. However this was probably only a technological rather than a commercial success. Shadrach Fox may have smelted iron with coke at Coalbrookdale in Shropshire in the 1690s, but only to make cannonballs and other cast iron products such as shells. However, in the peace after the Nine Years War, there was no demand for these.

Abraham Darby and his Successors

In 1707, Abraham Darby I patented a method of making cast iron pots. His pots were thinner and hence cheaper than those of his rivals. Needing a larger supply of pig iron he leased the blast furnace at Coalbrookdale in 1709. There, he made iron using coke, thus establishing the first successful business in Europe to do so. His products were all of cast iron, though his immediate successors attempted (with little commercial success) to fine this to bar iron.

Bar iron thus continued normally to be made with charcoal pig iron until the mid-1750s. In 1755 Abraham Darby II (with partners) opened a new coke-using furnace at Horsehay in Shropshire, and this was followed by others. These supplied coke pig iron to finery forges of the traditional kind for the production of bar iron. The reason for the delay remains controversial.

New Forge Processes

Schematic drawing of a puddling furnace.

It was only after this that economically viable means of converting pig iron to bar iron began to be devised. A process known as potting and stamping was devised in the 1760s and improved in the 1770s, and seems to have been widely adopted in the West Midlands from about 1785. However, this was largely replaced by Henry Cort's puddling process, patented in 1784, but probably only made to work with grey pig iron in about 1790. These processes permitted the great expansion in the production of iron that constitutes the Industrial Revolution for the iron industry.

In the early 19th century, Hall discovered that the addition of iron oxide to the charge of the puddling furnace caused a violent reaction, in which the pig iron was decarburised, this became known as 'wet puddling'. It was also found possible to produce steel by stopping the puddling process before decarburisation was complete.

Hot Blast

The efficiency of the blast furnace was improved by the change to hot blast, patented by James Beaumont Neilson in Scotland in 1828. This further reduced production costs. Within a few decades, the practice was to have a 'stove' as large as the furnace next to it into which the waste gas (containing CO) from the furnace was directed and burnt. The resultant heat was used to preheat the air blown into the furnace.

Industrial Steelmaking

Apart from some production of puddled steel, English steel continued to be made by the cementation process, sometimes followed by remelting to produce crucible steel. These were batch-based processes whose raw material was bar iron, particularly Swedish oregrounds iron.

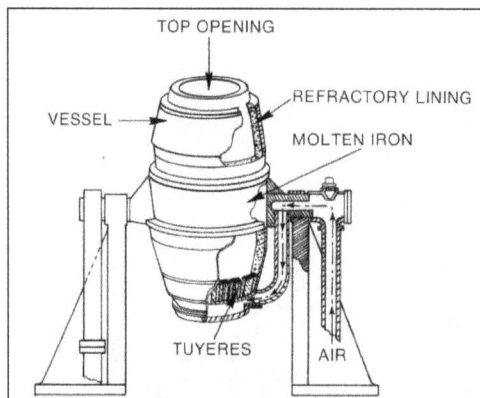

Schematic drawing of a Bessemer converter.

The problem of mass-producing cheap steel was solved in 1855 by Henry Bessemer, with the introduction of the Bessemer converter at his steelworks in Sheffield, England. (An early converter can still be seen at the city's Kelham Island Museum). In the Bessemer process, molten pig iron from the blast furnace was charged into a large crucible, and then air was blown through the molten iron from below, igniting the dissolved carbon from the coke. As the carbon burned off, the melting point of the mixture increased, but the heat from the burning carbon provided the extra energy needed to keep the mixture molten. After the carbon content in the melt had dropped to the desired level, the air draft was cut off: a typical Bessemer converter could convert a 25-ton batch of pig iron to steel in half an hour.

Finally, the basic oxygen process was introduced at the Voest-Alpine works in 1952; a modification of the basic Bessemer process, it lances oxygen from above the steel (instead of bubbling air from below), reducing the amount of nitrogen uptake into the steel. The basic oxygen process is used in all modern steelworks; the last Bessemer converter in the U.S. was retired in 1968. Furthermore, the last three decades have seen a massive increase in the mini-mill business, where scrap steel only is melted with an electric arc furnace. These mills only produced bar products at first, but have since expanded into flat and heavy products, once the exclusive domain of the integrated steelworks.

Until these 19th-century developments, steel was an expensive commodity and only used for a limited number of purposes where a particularly hard or flexible metal was needed, as in the cutting edges of tools and springs. The widespread availability of inexpensive steel powered the Second Industrial Revolution and modern society as we know it. Mild steel ultimately replaced wrought iron for almost all purposes, and wrought iron is no longer commercially produced. With minor exceptions, alloy steels only began to be made in the late 19th century. Stainless steel was developed on the eve of World War I and was not widely used until the 1920s.

METALLURGY OF IRON AND STEEL

The early application of iron to the manufacture of tools and weapons was possible because of the wide distribution of iron ores and the ease with which iron compounds in the ores could be reduced by carbon. For a long time, charcoal was the form of carbon used in the reduction process. The production and use of iron became much more widespread about 1620, when coke was introduced as the reducing agent. Coke is a form of carbon formed by heating coal in the absence of air to remove impurities.

The overall reaction for the production of iron in a blast furnace is as follows:

$$Fe_2O_3(s) + 3C(s) \xrightarrow{\Delta} 2Fe(l) + 3CO(g)$$

The actual reductant is CO, which reduces Fe_2O_3 to give Fe(l) and CO_2(g) the CO_2 is then reduced back to CO by reaction with excess carbon. As the ore, lime, and coke drop into the furnace any silicate minerals in the ore react with the lime to produce a low-melting mixture of calcium silicates called slag, which floats on top of the molten iron. Molten iron is then allowed to run out the bottom of the furnace, leaving the slag behind. Originally, the iron was collected in pools called pigs, which is the origin of the name pig iron.

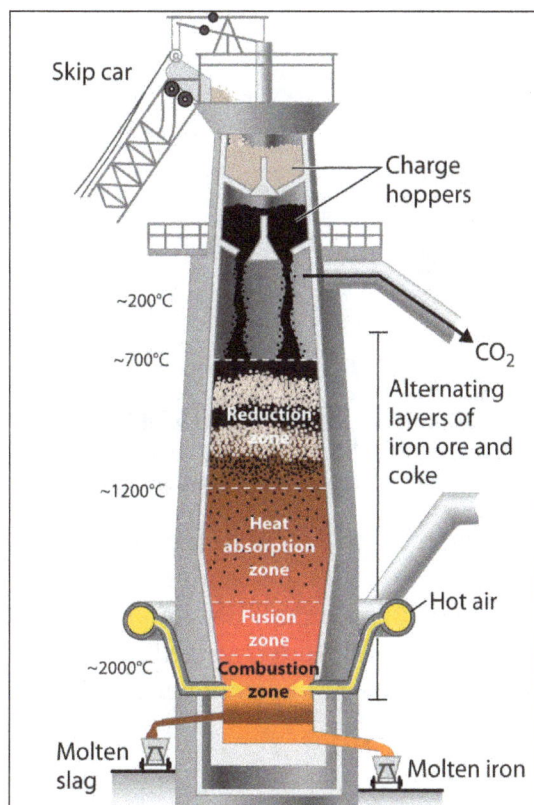

The above figure shows: A Blast Furnace for Converting Iron Oxides to Iron Metal. (a) The furnace is charged with alternating layers of iron ore (largely Fe_2O_3) and a mixture of coke (C) and limestone ($CaCO_3$). Blasting hot air into the mixture from the bottom causes it to ignite, producing CO and raising the temperature of the lower part of the blast furnace to about 2000 °C. As the CO that is formed initially rises, it reduces Fe_2O_3 to form CO_2 and elemental iron, which absorbs heat and melts as it falls into the hottest part of the furnace. Decomposition of $CaCO_3$ at high temperatures produces CaO (lime) and additional CO_2, which reacts with excess coke to form more CO.

The first step in the metallurgy of iron is usually roasting the ore (heating the ore in air) to remove water, decomposing carbonates into oxides, and converting sulfides into oxides. The oxides are then reduced in a blast furnace that is 80–100 feet high and about 25 feet in diameter in which the roasted ore, coke, and limestone (impure $CaCO_3$) are introduced continuously into the top. Molten iron and slag are withdrawn at the bottom. The entire stock in a furnace may weigh several hundred tons.

Near the bottom of a furnace are nozzles through which preheated air is blown into the furnace. As soon as the air enters, the coke in the region of the nozzles is oxidized to carbon dioxide, with the liberation of a great deal of heat. The hot carbon dioxide passes upward through the overlying layer of white-hot coke, where it is reduced to carbon monoxide:

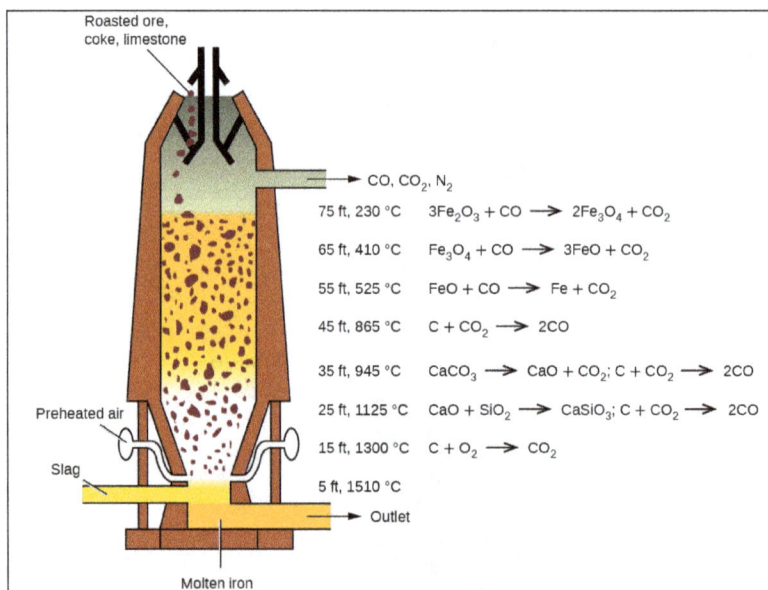

$$CO_2(g) + C(s) \rightarrow 2CO(g)$$

Within a blast furnace, different reactions occur in different temperature zones.
Carbon monoxide is generated in the hotter bottom regions and rises upward to reduce the
iron oxides to pure iron through a series of reactions that take place in the upper regions.

The carbon monoxide serves as the reducing agent in the upper regions of the furnace. The individual reactions are indicated The iron oxides are reduced in the upper region of the furnace. In the middle region, limestone (calcium carbonate) decomposes, and the resulting calcium oxide combines with silica and silicates in the ore to form slag. The slag is mostly calcium silicate and contains most of the commercially unimportant components of the ore:

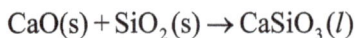

$$CaO(s) + SiO_2(s) \rightarrow CaSiO_3(l)$$

Just below the middle of the furnace, the temperature is high enough to melt both the iron and the slag. They collect in layers at the bottom of the furnace; the less dense slag floats on the iron and protects it from oxidation. Several times a day, the slag and molten iron are withdrawn from the furnace. The iron is transferred to casting machines or to a steelmaking plant.

Molten iron is shown being cast as steel.

Steel

Much of the iron produced is refined and converted into steel. Steel is made from iron by removing impurities and adding substances such as manganese, chromium, nickel, tungsten, molybdenum, and vanadium to produce alloys with properties that make the material suitable for specific uses. Most steels also contain small but definite percentages of carbon (0.04%–2.5%). However, a large part of the carbon contained in iron must be removed in the manufacture of steel; otherwise, the excess carbon would make the iron brittle. However, there is not just one substance called steel - they are a family of alloys of iron with carbon or various metals.

Impurities in the iron from the Blast Furnace include carbon, sulfur, phosphorus and silicon, which have to be removed.

- Removal of sulfur: Sulfur has to be removed first in a separate process. Magnesium powder is blown through the molten iron and the sulfur reacts with it to form magnesium sulfide. This forms a slag on top of the iron and can be removed.

    $$Mg + S \rightarrow MgS$$

- Removal of carbon: The still impure molten iron is mixed with scrap iron (from recycling) and oxygen is blown on to the mixture. The oxygen reacts with the remaining impurities to form various oxides. The carbon forms carbon monoxide. Since this is a gas it removes itself from the iron. This carbon monoxide can be cleaned and used as a fuel gas.

- Removal of other elements: Elements like phosphorus and silicon react with the oxygen to form acidic oxides. These are removed using quicklime (calcium oxide) which is added to the furnace during the oxygen blow. They react to form compounds such as calcium silicate or calcium phosphate which form a slag on top of the iron.

Special Steels			
	Iron mixed with	Special properties	Uses include
Stainless steel	Chromium and Nickel	Resists corrosion	cutlery, cooking utensils, kitchen sinks, industrial equipment for food and drink processing
Titanium steel	Titanium	Withstands high temperatures	gas turbines, spacecraft
Manganese steel	Manganese	Very hard	rock-breaking machinery, some railway track (e.g. points), military helmets

Cast iron has already been mentioned above. This section deals with the types of iron and steel which are produced as a result of the steel-making process.

- Wrought iron: If all the carbon is removed from the iron to give high purity iron, it is known as wrought iron. Wrought iron is quite soft and easily worked and has little structural strength. It was once used to make decorative gates and railings, but these days mild steel is normally used instead.

- Mild steel: Mild steel is iron containing up to about 0.25% of carbon. The presence of the carbon makes the steel stronger and harder than pure iron. The higher the percentage of carbon, the harder the steel becomes. Mild steel is used for lots of things - nails, wire, car bodies, ship building, girders and bridges amongst others.

- High carbon steel: High carbon steel contains up to about 1.5% of carbon. The presence of the extra carbon makes it very hard, but it also makes it more brittle. High carbon steel is used for cutting tools and masonry nails (nails designed to be driven into concrete blocks or brickwork without bending). High carbon steel tends to fracture rather than bend if mistreated.

- Special steels: These are iron alloyed with other metals.

STEELMAKING

Steel mill with two arc furnaces.

Steelmaking is the process of producing steel from iron ore and scrap. In steelmaking, impurities such as nitrogen, silicon, phosphorus, sulfur and excess carbon(most important impurity) are removed from the sourced iron, and alloying elements such as manganese, nickel, chromium, carbon and vanadium are added to produce different grades of steel. Limiting dissolved gases such as nitrogen and oxygen and entrained impurities (termed "inclusions") in the steel is also important to ensure the quality of the products cast from the liquid steel.

Steelmaking has existed for millennia, but it was not commercialized on a massive scale until late 19th century. The ancient craft process of steelmaking was the crucible process. In the 1850s and 1860s, the Bessemer process and the Siemens-Martin process turned steelmaking into a heavy industry. Today there are two major commercial processes for making steel, namely basic oxygen steelmaking, which has liquid pig-iron from the blast furnace and scrap steel as the main feed materials, and electric arc furnace (EAF) steelmaking, which uses scrap steel or direct reduced iron (DRI) as the main feed materials. Oxygen steelmaking is fuelled predominantly by the exothermic nature of the reactions inside the vessel; in contrast, in EAF steelmaking, electrical energy is used to melt the solid scrap and DRI materials. In recent times, EAF steelmaking technology has evolved closer to oxygen steelmaking as more chemical energy is introduced into the process.

Modern Processes

Modern steelmaking processes can be divided into two categories:

- Primary steel making.

- Secondary steelmaking.

Primary steelmaking involves converting liquid iron from a blast furnace and steel scrap into steel via basic oxygen steelmaking, or melting scrap steel or direct reduced iron (DRI) in an electric arc furnace.

Secondary steelmaking involves refining of the crude steel before casting and the various operations are normally carried out in ladles. In secondary metallurgy, alloying agents are added, dissolved gases in the steel are lowered, and inclusions are removed or altered chemically to ensure that high-quality steel is produced after casting. Primary and secondary steel making process are described below:

Primary Steelmaking

Basic oxygen steelmaking is a method of primary steelmaking in which carbon-rich molten pig iron is converted into steel. Blowing oxygen through molten pig iron lowers the carbon content of the alloy and changes it into steel. The process is known as *basic* due to the chemical nature of the refractories—calcium oxide and magnesium oxide—that line the vessel to withstand the high temperature and corrosive nature of the molten metal and slag in the vessel. The slag chemistry of the process is also controlled to ensure that impurities such as silicon and phosphorus are removed from the metal.

The process was developed in 1948 by Robert Durrer, using a refinement of the Bessemer converter where blowing of air is replaced with blowing oxygen. It reduced capital cost of the plants, time of smelting, and increased labor productivity. Between 1920 and 2000, labour requirements in the industry decreased by a factor of 1000, from more than 3 man-hours per tonne to just 0.003 man hour. The vast majority of steel manufactured in the world is produced using the basic oxygen furnace in 2011, it accounted for 70% of global steel output. Modern furnaces will take a charge of iron of up to 350 tons and convert it into steel in less than 40 minutes compared to 10–12 hours in an open hearth furnace.

Electric arc furnace steelmaking is the manufacture of steel from scrap or direct reduced iron melted by electric arcs. In an electric arc furnace, a batch of steel ("heat") may be started by loading scrap or direct reduced iron into the furnace, sometimes with a "hot heel" (molten steel from a previous heat). Gas burners may be used to assist with the melt down of the scrap pile in the furnace. As in basic oxygen steelmaking, fluxes are also added to protect the lining of the vessel and help improve the removal of impurities. Electric arc furnace steelmaking typically uses furnaces of capacity around 100 tonnes that produce steel every 40 to 50 minutes for further processing.

By-product gases from the steel making process can be used to generate electricity through the use of reciprocating gas engines/gas turbines. These green house gases are produced by burning fossil fuels contributing to global warming.

Secondary Steelmaking

Secondary steelmaking is most commonly performed in ladles. Some of the operations performed in ladles include de-oxidation (or "killing"), vacuum degassing, alloy addition, inclusion removal, inclusion chemistry modification, de-sulphurisation, and homogenisation. It is now common to perform ladle metallurgical operations in gas-stirred ladles with electric arc heating in the lid of the furnace. Tight control of ladle metallurgy is associated with producing high grades of steel in which the tolerances in chemistry and consistency are narrow.

Hlsarna Steelmaking

The Hlsarna ironmaking process is a process in which iron ore is processed almost directly into liquid iron or hot metal. The process is based around a type of blast furnace called a *cyclone converter furnace*, which makes it possible to skip the process of manufacturing pig iron pellets that is necessary for the basic oxygen steelmaking process. Without the necessity of this preparatory step, the Hlsarna process is more energy-efficient and has a lower carbon footprint than traditional steelmaking processes.

PRIMARY METAL PRODUCTION

Primary Aluminum Production

The primary ore of aluminum is bauxite, a mixture of hydrated aluminum oxides:

- Gibbsite: $Al(OH)_3$ (most extractable form).

- Boehmite: $\gamma AlO \cdot OH$ (less extractable than Gibbsite).

- Diaspore: $\alpha AlO \cdot OH$ (difficult to extract).

It is formed by weathering of aluminum-bearing rocks by rainwater, and so bauxite deposits tend to be found in areas that are now, or were in the past, tropical high-rainfall areas. Aluminum is contained in many minerals, but bauxite is the preferred ore because it has the highest aluminum oxide content and is therefore the cheapest to process. Although aluminum is an extremely common element on earth, it was not practical to produce aluminum metal at a reasonable cost until two breakthroughs had been made: a method for producing purified aluminum oxide from bauxite (the Bayer process), and a method for converting aluminum oxide to metallic aluminum (the Hall-Heroult process).

Reactions

Bayer Process

The Bayer process consists of three steps:

1. Dissolving the aluminum hydroxides from bauxite with hot (250° C) sodium hydroxide solution.

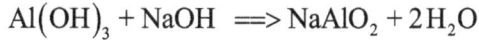

$$Al(OH)_3 + NaOH \implies NaAlO_2 + 2H_2O$$
$$AlO \bullet OH + NaOH \implies NaAlO_2 + H_2O$$

As an undesirable side reaction, Kaolinite clay dissolves, reacts with alumina and caustic, and precipitates as "red mud":

$$5Al_2Si_2O_5(OH)_4 + 2Al(OH)_3 + 12NaOH \implies 2Na_6Al_6Si_5O_{17}(OH)_{10} + 10H_2O$$

2. Crystallizing purified aluminum hydroxide by addition of seed crystals:

$$NaAlO_2 + 2H_2O \implies Al(OH)_3 + NaOH$$

3. Calcining aluminum hydroxide to aluminum oxide:

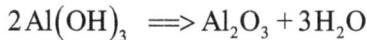

$$2Al(OH)_3 \implies Al_2O_3 + 3H_2O$$

Hall-Heroult Process

This process consists of dissolving aluminum hydroxide in molten cryolite (Na_3AlF_6) and electrolyzing the molten salt at 1.5 - 1.7 volts with a consumable carbon anode, to produce molten metallic aluminum and carbon dioxide:

$$2Al_2O_3 + 3C \implies 4Al + 3CO_2$$

Alumina Production using the Bayer Process

The Bayer process is the primary method for producing purified alumina (Al_2O_3) from bauxite. It is a hydrometallurgical process, using concentrated alkali solution to dissolve hydrated aluminum oxide from bauxite in an autoclave at elevated temperatures, followed by solid-liquid separation and precipitation and calcination of purified alumina. Processing before the Bayer process is minimal, usually being little more than a rotary breaker to remove coarse rock from the bauxite.

Dissolution

The first step is to selectively dissolve the hydrated aluminum oxides from the bauxite, while leaving behind the bulk of the silicate minerals and other impurities:

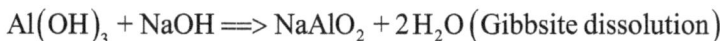

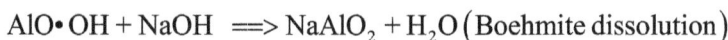

$$Al(OH)_3 + NaOH \implies NaAlO_2 + 2H_2O \text{ (Gibbsite dissolution)}$$
$$AlO \bullet OH + NaOH \implies NaAlO_2 + H_2O \text{ (Boehmite dissolution)}$$

In most bauxites, Gibbsite and Boehmite are the main aluminum-bearing phases. Gibbsite dissolves most readily in the alkali solution, and Boehmite dissolves with some difficulty. Diaspore, the third form of hydrated aluminum oxide, is difficult to dissolve and is generally not recovered in the Bayer process. When Diaspore is an important phase in a bauxite, it can only be dissolved if lime is added to the digester, and the digestion temperature is increased to about 300° C.

The conditions for efficient dissolution of the aluminum oxides are as follows:

- Caustic Soda ($Na2CO3$ equivalent): 150-255 grams/liter (Gibbsite), 205-445 grams/liter (Boehmite).

- Temperature: 100 - 150 °C (Gibbsite), 200 - 315 °C (Boehmite).

- Autoclave Pressure: 56-70 psi (Gibbsite), 280-1500 psi (Boehmite).

- Feed Slurry Solids Content: 50% wt. Bauxite.

- Digestion Time: 30 minutes - 4 hours.

Note that the temperatures are well above the boiling point of water at atmospheric pressure. An autoclave is therefore needed for the digester to raise the boiling point high enough.

Alkali is normally purchased as caustic soda (Na_2CO_3), but it needs to be converted to NaOH in order to work properly. The conversion is done by causticizing with lime ($Ca(OH)_2$):

$$Na_2CO_3 + Ca(OH)_2 \implies 2NaOH + CaCO_3$$

This has the added benefit that any phosphate dissolved from the bauxite will be precipitated by the lime as $Ca_5(PO_4)_3OH$.

If Kaolinite or other clays are present, they also dissolve in the alkali solution. For Kaolinite, the dissolution reaction is:

$$5Al_2Si_2O_5(OH)_4 + 2Al(OH)_3 + 12NaOH \implies 2Na_6Al_6Si_5O_{17}(OH)_{10} + 10H_2O$$

The precipitate from this reaction is a material referred to as "red mud", which is a fine material that is difficult to remove from the solution. Note that the formation of red mud precipitate consumes dissolved aluminum, and so the red mud represents an aluminum loss.

Flash Cooling

The temperature of the slurry is reduced by progressively reducing its pressure in a series of tanks, allowing a portion of the liquid phase to flash to steam in each tank.

The steam goes to heat exchangers, where it preheats leaching solution returning to the digester. This recycling of heat is critical for keeping the fuel costs for the process as low as possible.

Once the slurry temperature is reduced to about 100 °C, it does not need to be kept in pressure vessels any longer. It is then diluted to about 3% solids for solid/liquid separation.

Solid-liquid Separation

Once the bauxite has been digested, the slurry contains one liquid phase and two solid components:

- Caustic liquid solution with the dissolved aluminum.

- Undissolved coarse material (sand).

- Precipitated fines (red mud).

The solids must be separated from the liquid phase so that they will not contaminate the alumina product when it is precipitated from solution.

The coarse sands, which are mainly undissolved silicate grains from the bauxite, settle from suspension very rapidly and are easily removed by gravity classifiers or hydrocyclones. They are washed in rake or spiral classifiers to remove dissolved alumina from them, and then disposed of.

The fine red mud has a very slow settling rate, and is removed by large-diameter thickeners after adding a flocculant to increase the red mud settling rate.

The liquid overflow from the thickener still contains about 50-100 mg of solids per liter. The remaining solid particles are removed from this stream by filter presses.

The red mud underflow from the thickener is 15-25% solids, and contains a great deal of dissolved aluminum that needs to be recovered. This is done using a counter-current decantation (CCD) mud-washing circuit. This is a series of thickeners can be used to remove the liquid from the solids with a minimum of dilution by washwater. The effect of the arrangement shown is that the solids travel down the series of thickeners in one direction, while the wash solution travels up the series in the opposite direction. In each tank, the solids are being washed by a solution with a lower metal concentration than in the previous tank, until it reaches the final tank where the wash solution contains no dissolved metal at all. The final solids product is often either centrifuged or filtered to remove the last bit of liquid with dissolved metals.

The red mud is a disposal problem for the aluminum industry. No practical uses have been found for the material yet, and so it must be pumped into impoundment areas.

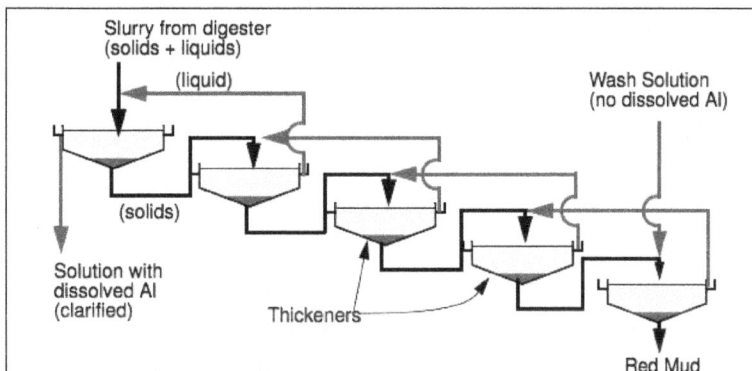

Series of thickeners, being used for continuous countercurrent decantation.

Precipitation

Once the liquid solution has been clarified and cooled, it is a supersaturated solution containing about 100-175 grams of dissolved alumina per liter. Since $Al(OH)_3$ does not crystallize easily on its own, it is precipitated by adding seed crystals in long-residence-time agitated precipitator tanks. Crystallization takes many hours, and so the tanks must be large (24-36 feet in diameter, 60-80 feet tall).

The seed crystals of $Al(OH)_3$ are typically added at a rate of 40-300 grams/liter, and are 50-95% passing 74 μm. The pregnant liquor contains 100-175 grams of alumina per liter, and about 45-75 grams per liter are precipitated out on the seed crystals.

When the slurry exits the precipitators, the crystals that have reached the target size are removed by a classifier, washed, and filtered. Crystals that are smaller than the target size are concentrated by a thickener, and recirculated back to the precipitators to act as seed crystals. The spent liquor is reheated, recausticized, and sent back to the digester.

Calcination

The product from the precipitation process is wet crystals of $Al(OH)_3$. This must be dried and converted to Al_2O_3 before it can be used as feed to the Hall-Heroult process. This conversion is done by heating to approximately 1300 - 1500°C (fine, "floury" alumina requires the higher temperatures).

Calcination can be carried out using rotary kilns, or fluid-bed calciners. The fluid-bed units are reported to use 1500 BTU per lb of alumina produced, compared to 2200 BTU per lb for rotary kilns.

Hall-Heroult Process

Many reactive metals, such as magnesium and sodium, can be produced by electrolyzing a molten chloride salt of the metal. Unfortunately, aluminum chloride can not be electrolyzed easily because it sublimates rather than melting. Even if the pressure is increased enough to allow it to have a molten phase, molten aluminum chloride is an electrical insulator and so cannot be used as an electrolyte. Because of this, aluminum was produced by an expensive sodium reduction process until a suitable electrolyte was found.

In 1886, Hall (in the U. S.) and Heroult (in France) independently developed processes using cryolite (Na_3AlF_6) as a molten salt electrolyte for producing aluminum. Cryolite is electrically conductive, and dissolves alumina.

The electrolysis cell used is shown in figure. Both the anodes and the cathode are made of carbon. The anodes are gradually consumed by the oxygen that migrates to the anodes, and so the overall electrolysis reaction is:

$$2\,Al_2O_3 + 3C \implies 4\,Al + 3\,CO_2$$

The theoretical voltage for this reaction to occur is 1.15 volts, but due to anode overvoltages the potential in actual practice is 1.5-1.7 volts. In order to overcome the electrical resistance of the electrodes, conductors, and containers, the typical operating voltage is 4-5 volts.

Operating conditions for an aluminum electrolysis cell are:

- Temperature: 935-975 °C.

- Alumina content of electrolyte: 2-6% .

- Cell voltage: 4-5 volts.

- Faraday efficiency: 85-90%.

Additives to cryolite bath:

- AlF_3 (improves Faraday efficiency).

- CaF_2, LiF, MgF_2 (reduce freezing point of electrolyte).

Raw material and power usage per ton of Al produced:

- Al_2O_3: 1.90-1.95.

- Electrolyte: 0.04-0.06.

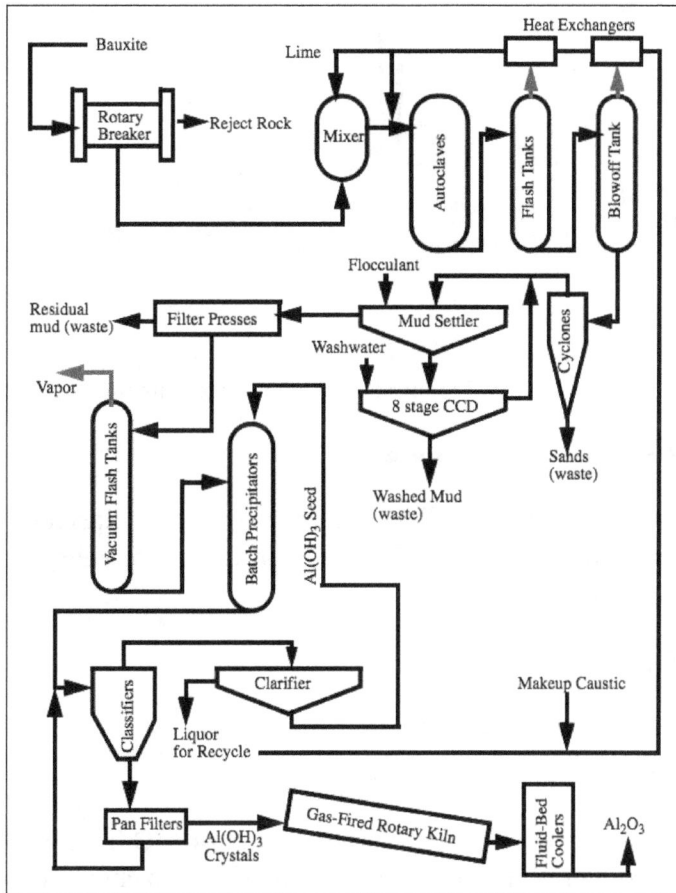

Flowsheet for a Bayer process plant, based on the
Kaiser Aluminum plant in Gramercy, Louisiana.

- Anode carbon: 0.43-0.50.

- Power, Kw-hr: 13,000-16,000.

Wastes produced in aluminum electrolysis:

- HF, CF_4 , and other fluorocarbons (from anode reactions).

- CO_2.

- "Salt cake" (spent electrolyte, metal oxides, and entrained metallic aluminum droplets).

Note the tremendous power consumption of aluminum production. The electrolysis must supply three electrons for every atom of metallic aluminum, and so very high currents are needed to produce aluminum at a reasonable rate. Because of this, electric power is the single largest cost in aluminum production, and so aluminum smelters are typically located in areas where electric power is inexpensive, generally near major hydropower sites.

The high power consumption of primary aluminum production also encourages aluminum recycling, which requires much less electric power.

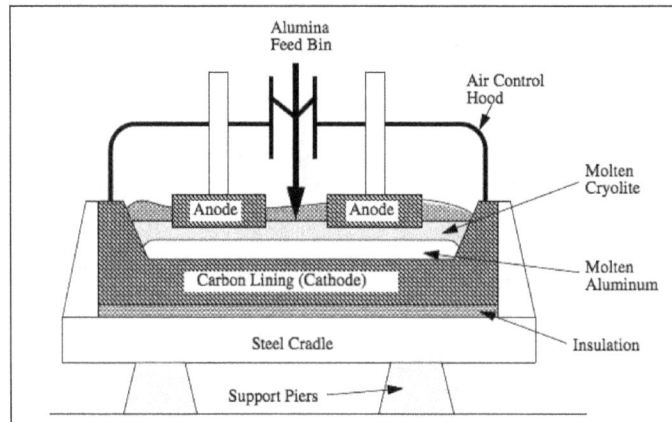

Electrolysis cell for production of metallic aluminum.

Other Aluminum Production Processes

Because of the high energy costs of the Hall-Heroult process, there has been considerable interest in finding alternative aluminum production routes that are more energy-efficient. A few of the more promising approaches are as follows:

Carbothermal Reduction

The reaction of aluminum oxide with carbon:

$$Al_2O_3 + 3C \implies 2Al + 3CO$$

can theoretically occur at temperatures greater than about 2000°C. This would provide a relatively inexpensive method for producing aluminum, because carbon is a cheap source of reducing power. However, there are several side reactions that form oxycarbides, aluminum carbide, and volatile monovalent aluminum compounds, and these side reactions make the overall carbothermic reduction much harder to operate and control.

An example of a carbothermic process is one patented by Alcoa in 1976. This is a two-step process, using two baths of molten aluminum oxide. The first step is:

$$2Al_2O_3 (l) + 9C \implies Al_4C_3 (l) + 6CO (2050°C)$$

The Al_4C_3 produced is then sent to the second bath, where the reaction is:

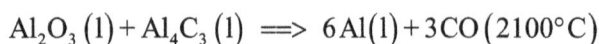

$$Al_2O_3 (l) + Al_4C_3 (l) \implies 6Al(l) + 3CO (2100°C)$$

to produce metallic aluminum.

Toth Process

This process is based on the reduction of AlCl3 by manganese, and consists of four steps. The first step is carbochlorination of alumina, where the alumina is provided as kaolinite clay:

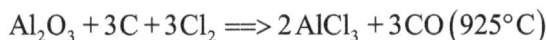

$$Al_2O_3 + 3C + 3Cl_2 ==> 2AlCl_3 + 3CO \, (925°C)$$

The aluminum chloride is then reacted with manganese to make metallic aluminum:

$$2AlCl_3 + 3Mn ==> 2Al + 3MnCl_2 \, (230°C)$$

The chlorine is then regenerated by reacting the manganese with oxygen:

$$2MnCl_2 + O_2 ==> 2MnO + 2Cl_2 \, (600°C)$$

and the manganese oxide is reduced with carbon to regenerate metallic manganese:

$$MnO + C ==> Mn + CO \, (1750°C)$$

While this uses only about 5% as much electricity as the Hall-Heroult process, it consumes as much carbon as the carbothermal process. It is unlikely to ever be used industrially, because of the great difficulty in regenerating the manganese.

Alcan Process

This is based on the production of monovalent aluminum as AlCl, which is highly volatile. The Alcan process first melts the aluminum-bearing feed in an electric furnace, and reduces it with carbon to produce a liquid alloy of iron, silicon, and aluminum. This alloy is then reacted with AlCl3 at about 1300 °C, which produces AlCl:

$$2Al(\text{from alloy}) + AlCl_3 ==> 3AlCl(g)$$

The AlCl vapor is then condensed by a shower of liquid aluminum, and the AlCl disproportionates back into aluminum metal and $AlCl_3$:

$$3AlCl ==> AlCl_3 + 2Al$$

This process does not require purified alumina, and so it can use materials such as raw bauxite or clay as feedstock, skipping the need for the Bayer process. Unfortunately, other volatile halides are removed from the alloy along with the AlCl, resulting in an impure product, and the chlorides produce a very corrosive environment, and so the overall process was less economical than the Hall-Heroult process.

Aluminum Chloride Electrolysis

There has long been interest in direct electrolysis of aluminum chloride to produce metal, according to the reaction:

$$2AlCl_3 ==> 2Al + 3Cl_2$$

The attractive features of this reaction are:

- Considerably lower working temperature than cryolite electrolysis.

- Higher current densities can be achieved before the "anode effect" halts production.

- Consumable carbon electrodes are not needed, which results in lower operating costs and less pollution.

The drawbacks are:

- Pure aluminum chloride sublimates at atmospheric pressure, and so it either needs to be pressurized to produce a liquid phase, or it needs to be dissolved in another salt, such as NaCl or KCl.

- Pure aluminum chloride is an electrical insulator, and so it needs to be dissolved in a conductive salt (again, NaCl or KCl can be used).

- Aluminum chloride is produced from pure alumina, which is made by the Bayer process. The cost of converting the alumina to aluminum chloride is a considerable expense.

A process of this type was developed by Alcoa, and was considered for industrial use, although it apparently has not yet been used on larger than a pilot scale.

The Alcoa process required high-purity electrolyte from which all oxygen had been excluded, which is why the two filtration stages and the inert-gas-purged evaporator were needed. A special electrolytic cell was used that was lined with oxysilicon nitride to prevent corrosion. The electrodes were graphite. Since chlorine was being produced at the anode rather than oxygen, the graphite anode was not consumed. A typical composition for the electrolyte was 5% $AlCl_3$, 53% NaCl, 40% LiCl, 0.5% $MgCl_2$, 0.5% KCl, and 1% $CaCl_2$.

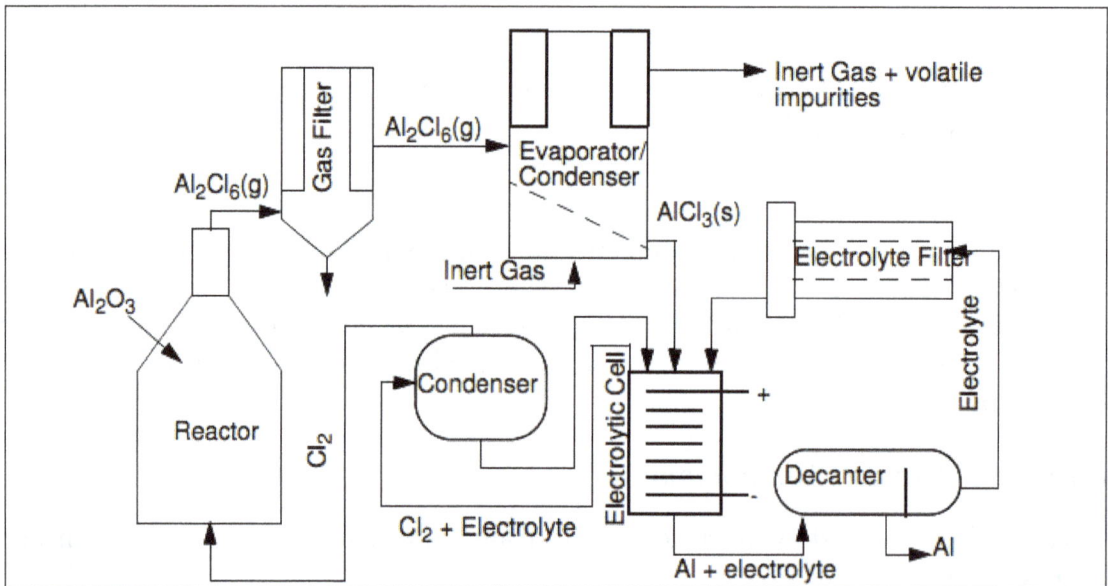

Flowsheet of the Alcoa aluminum chloride electrolysis process. "Electrolyte" signifies a mixture of alkaline chlorides, usually NaCl, LiCl, $MgCl_2$, KCl, AlC_{l3}, and $CaCl_2$.

Titanium Production

Titanium is quite expensive for a structural metal, at \$3 - \$4 per lb for titanium sponge. Its cost is not due to scarcity (it is quite common, being the 9th most common element in the Earth's crust), it is expensive because of the difficulty in reducing titanium oxides to metal. Since the primary use of the metal is for aerospace, the demand fluctuates greatly depending on aircraft orders in a given year. U. S. consumption has varied between 18 million tons/year and 28 million tons/year since 1998.

Titanium is in high demand for aircraft because it has a better strength-to-weight ratio than aluminum or steel. It is also desirable for marine applications because it is almost completely corrosion-resistant due to the extreme chemical stability of the thin oxide layer that it forms on its surface.

Metal production only accounts for about 3% of all the titanium minerals mined. A much greater quantity of titanium is used for producing titanium dioxide pigment, with over 1.3 billion tons of purified TiO_2 produced in the U. S. annually. It is in great demand as a pigment because in purified form it is white, and its high refractive index gives it excellent hiding power. It is also non-toxic, and has completely replaced the toxic lead-based pigments that were used previously in almost all paint.

Sources of Titanium

Titanium can be extracted economically from:

- Rutile (TiO_2).

- Ilmenite ($FeTiO_3$).

- Leucoxine (rutile contaminated with iron oxides, formed by alteration of ilmenite).

Both rutile and ilmenite are hard, dense minerals that resist weathering, and as a result they tend to concentrate in placer sand deposits where they are mined by dredging, and separated from other valuable minerals by a combination of density, magnetic, and electrostatic separations.

Purification and Metal Production

A typical flowsheet for producing titanium from rutile is shown in figure. The basic process consists of:

1. Roast the impure titanium dioxide with carbon in the presence of chlorine:

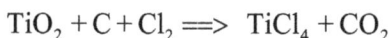

$$TiO_2 + C + Cl_2 \implies TiCl_4 + CO_2$$

If the titanium dioxide is reacted with carbon alone, it will produce highly-refractory TiC, which is extremely difficult to process further. The chlorine is needed to convert the titanium to the highly-volatile chloride, which vaporizes readily.

2. Distill the titanium tetrachloride ($TiCl_4$) to remove chlorides of iron and other metals. Titanium tetrachloride is liquid at room temperature, and takes relatively little energy to distill.

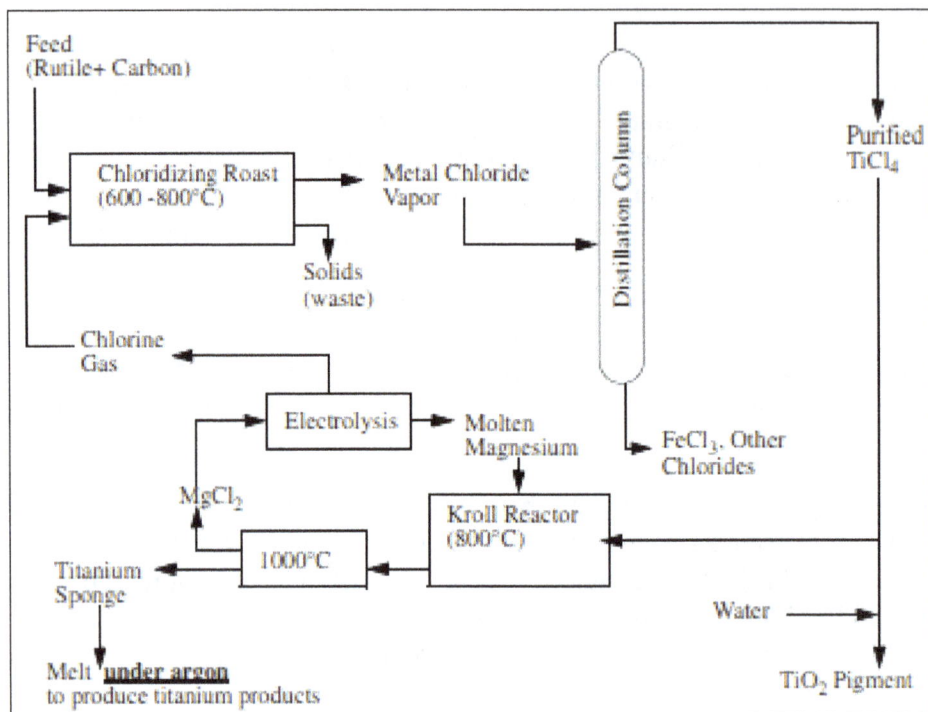

Flowsheet for producing titanium sponge from rutile.
To produce titanium dioxide pigment, the purified TiCl$_4$ can be reacted
with water to produce a white, finely-divided, iron-free product.

3. If pigment is the desired product, react the liquid titanium tetrachloride with water to produce high-purity fine titanium dioxide:

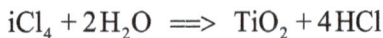

$$iCl_4 + 2H_2O \implies TiO_2 + 4HCl$$

To produce metal, the titanium tetrachloride is reacted with molten magnesium metal, in the Kroll process:

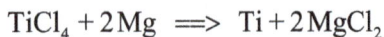

$$TiCl_4 + 2Mg \implies Ti + 2MgCl_2$$

This results in a porous sponge of titanium metal, filled with magnesium chloride.

4. The product from the Kroll reaction is heated to melt the magnesium chloride, and separate it from the titanium sponge. The molten magnesium chloride is then electrolyzed to produce magnesium metal and chlorine gas, which are recycled back into the process:

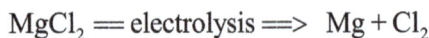

$$MgCl_2 \text{ } \text{===electrolysis} \implies Mg + Cl_2$$

5. The titanium sponge is then melted under an inert atmosphere to form into ingots for further fabrication.

There are a number of important features that contribute to the high cost of titanium production:

- The chloridizing roast gases are extremely corrosive and toxic, and must be completely contained. Maintaining gas-tightness in high-temperature reactors is difficult and expensive.

- Titanium tetrachloride is highly reactive with water and oxygen, and so all air and moisture must be rigidly excluded from all process stages.

- Molten titanium metal will not only react with oxygen, it will burn in nitrogen, so all melting must be carried out under an inert atmosphere (usually argon).

References

- Chemicalpropertiesofmetals, sthash, Metals, Chemistry, study-zone, funscience.in, Retrieved 1 January, 2019

- Sanderson, Katharine (2006-11-15). "Sharpest cut from nanotube sword: Carbon nanotech may have given swords of Damascus their edge". Nature. Archived from the original on 2006-11-19. Retrieved 2006-11-17

- Metallurgy, science: britannica.com, Retrieved 2 February, 2019

- Ghosh, Ahindra. (December 13, 2000). Secondary Steelmaking: Principles and Applications (1st ed.). Boca Raton, Fla.: CRC Press. ISBN 9780849302640. LCCN 00060865. OCLC 664116613

- A-Metallurgy-of-Iron-and-Steel, A-The-Transition-Elements, A-General-Chemistry, General-Chemistry: chem.libretexts.org, Retrieved 3 March, 2019

- Miller, Duncan E.; Der Merwe, N.J. Van (1994). "Early Metal Working in Sub-Saharan Africa: A Review of Recent Research". Journal of African History. 35: 1–36. Doi:10.1017/s0021853700025949

- Primary-Metals, kawatra, faculty, chem-eng, chem.mtu.edu, Retrieved 4 April, 2019

Extractive Metallurgy

The science of extracting valuable metals from their ore and refining them to achieve their purest form is referred to as extractive metallurgy. Mineral processing, smelting, hydrometallurgy, pyrometallurgy, electrometallurgy, etc. are some the concepts related to it. This chapter discusses in detail the concepts related to extractive metallurgy.

Extractive metallurgy refers to the different processes used to extract valuable metals from mined ores. Following separation and concentration by mineral processing, metallic minerals are subjected to extractive metallurgy, in which their metallic elements are extracted from chemical compound form and refined of impurities.

Metallic compounds are frequently rather complex mixtures (those treated commercially are for the most part sulfides, oxides, carbonates, arsenides, or silicates), and they are not often types that permit extraction of the metal by simple, economical processes. Consequently, before extractive metallurgy can effect the separation of metallic elements from the other constituents of a compound, it must often convert the compound into a type that can be more readily treated. Common practice is to convert metallic sulfides to oxides, sulfates, or chlorides; oxides to sulfates or chlorides; and carbonates to oxides. The processes that accomplish all this can be categorized as either pyrometallurgy or hydrometallurgy. Pyrometallurgy involves heating operations such as roasting, in which compounds are converted at temperatures just below their melting points, and smelting, in which all the constituents of an ore or concentrate are completely melted and separated into two liquid layers, one containing the valuable metals and the other the waste rock. Hydrometallurgy consists of such operations as leaching, in which metallic compounds are selectively dissolved from an ore by an aqueous solvent, and electrowinning, in which metallic ions are deposited onto an electrode by an electric current passed through the solution.

Extraction is often followed by refining, in which the level of impurities is brought lower or controlled by pyrometallurgical, electrolytic, or chemical means. Pyrometallurgical refining usually consists of the oxidizing of impurities in a high-temperature liquid bath. Electrolysis is the dissolving of metal from one electrode of an electrolytic cell and its deposition in a purer form onto the other electrode. Chemical refining involves either the condensation of metal from a vapour or the selective precipitation of metal from an aqueous solution.

The processes to be used in extraction and refining are selected to fit into an overall pattern, with the product from the first process becoming the feed material of the second process, and so on. It is quite common for hydrometallurgical, pyrometallurgical, and electrolytic processes to be used one after another in the treatment of a single metal. The choices depend on several conditions. One of these is that certain types of metallic compounds lend themselves to easiest extraction by certain methods; for example- oxides and sulfates are readily dissolved in leach solutions, while sulfides are only slightly soluble. Another condition is the degree of purity, which can vary from one type of extraction to another. Zinc production illustrates this, in that zinc metal produced by pyrometallurgical retort or blast-furnace operations is 98 percent pure, with traces of lead, iron, and cadmium. This is adequate for galvanizing, but the preferred purity for die-casting (99.99 percent) must be obtained hydrometallurgically, from the electrolysis of a zinc sulfate solution. Also to be considered in choosing a processing method is the recovery of particular impurities that may have value themselves as by-products. One example is copper refining: the pyrometallurgical refining of blister copper removes many impurities, but it does not recover or remove silver or gold; the precious metals are recovered, however, by subsequent electrolytic refining.

MINERAL PROCESSING

Mineral processing is the art of treating crude ores and mineral products in order to separate the valuable minerals from the waste rock, or gangue. It is the first process that most ores undergo after mining in order to provide a more concentrated material for the procedures of extractive metallurgy. The primary operations are comminution and concentration, but there are other important operations in a modern mineral processing plant, including sampling and analysis and dewatering.

Sampling and Analysis

Routine sampling and analysis of the raw material being processed are undertaken in order to acquire information necessary for the economic appraisal of ores and concentrates. In addition, modern plants have fully automatic control systems that conduct in-stream analysis of the material as it is being processed and make adjustments at any stage in order to produce the richest possible concentrate at the lowest possible operating cost.

Sampling

Sampling is the removal from a given lot of material a portion that is representative of the whole yet of convenient size for analysis. It is done either by hand or by machine. Hand sampling is usually expensive, slow, and inaccurate, so that it is generally applied only where the material is not suitable for machine sampling (slimy ore, for example) or where machinery is either not available or too expensive to install.

Many different sampling devices are available, including shovels, pipe samplers, and automatic machine samplers. For these sampling machines to provide an accurate representation of the whole lot, the quantity of a single sample, the total number of samples, and the kind of samples taken are of decisive importance. A number of mathematical sampling models have been devised in order to arrive at the appropriate criteria for sampling.

Analysis

After one or more samples are taken from an amount of ore passing through a material stream such as a conveyor belt, the samples are reduced to quantities suitable for further analysis. Analytical methods include chemical, mineralogical, and particle size.

Chemical Analysis

Even before the 16th century, comprehensive schemes of assaying (measuring the value of) ores were known, using procedures that do not differ materially from those employed in modern times. Although conventional methods of chemical analysis are used today to detect and estimate quantities of elements in ores and minerals, they are slow and not sufficiently accurate, particularly at low concentrations, to be entirely suitable for process control. As a consequence, to achieve greater efficiency, sophisticated analytical instrumentation is being used to an increasing extent.

In emission spectroscopy, an electric discharge is established between a pair of electrodes, one of which is made of the material being analyzed. The electric discharge vaporizes a portion of the sample and excites the elements in the sample to emit characteristic spectra. Detection and measurement of the wavelengths and intensities of the emission spectra reveal the identities and concentrations of the elements in the sample.

In X-ray fluorescence spectroscopy, a sample bombarded with X rays gives off fluorescent X-radiation of wavelengths characteristic of its elements. The amount of emitted X-radiation is related to the concentration of individual elements in the sample. The sensitivity and precision of this method are poor for elements of low atomic number (i.e., few protons in the nucleus, such as boron and beryllium), but for slags, ores, sinters, and pellets where the majority of the elements are in the higher atomic number range, as in the case of gold and lead, the method has been generally suitable.

Mineralogical Analysis

A successful separation of a valuable mineral from its ore can be determined by heavy-liquid testing, in which a single-sized fraction of a ground ore is suspended in a liquid of high specific gravity. Particles of less density than the liquid remain afloat, while denser particles sink. Several different fractions of particles with the same density (and hence, similar composition) can be produced, and the valuable mineral components can then be determined by chemical analysis or by microscopic analysis of polished sections.

Size Analysis

Coarsely ground minerals can be classified according to size by running them through special sieves or screens, for which various national and international standards have been accepted. One old standard (now obsolete) was the Tyler Series, in which wire screens were identified by mesh size, as measured in wires or openings per inch. Modern standards now classify sieves according to the size of the aperture, as measured in millimetres or micrometres (10^{-6} metre).

Mineral particles smaller than 50 micrometres can be classified by different optical measurement methods, which employ light or laser beams of various frequencies.

Comminution

In order to separate the valuable components of an ore from the waste rock, the minerals must be liberated from their interlocked state physically by comminution. As a rule, comminution begins by crushing the ore to below a certain size and finishes by grinding it into powder, the ultimate fineness of which depends on the fineness of dissemination of the desired mineral.

In primitive times, crushers were small, hand-operated pestles and mortars, and grinding was done by millstones turned by men, horses, or waterpower. Today, these processes are carried out in mechanized crushers and mills. Whereas crushing is done mostly under dry conditions, grinding mills can be operated both dry and wet, with wet grinding being predominant.

Crushing

Some ores occur in nature as mixtures of discrete mineral particles, such as gold in gravel beds and streams and diamonds in mines. These mixtures require little or no crushing, since the valuables are recoverable using other techniques (breaking up placer material in log washers, for instance). Most ores however, are made up of hard, tough rock masses that must be crushed before the valuable minerals can be released.

In order to produce a crushed material suitable for use as mill feed (100 percent of the pieces must be less than 10 to 14 millimetres, or 0.4 to 0.6 inch, in diameter), crushing is done in stages. In the primary stage, the devices used are mostly jaw crushers with openings as wide as two metres. These crush the ore to less than 150 millimetres, which is a suitable size to serve as feed for the secondary crushing stage. In this stage, the ore is crushed in cone crushers to less than 10 to 15 millimetres. This material is the feed for the grinding mill.

Grinding

In this process stage, the crushed material can be further disintegrated in a cylinder mill, which is a cylindrical container built to varying length-to-diameter ratios, mounted with the axis substantially horizontal, and partially filled with grinding bodies (e.g., flint stones, iron or steel balls) that are caused to tumble, under the influence of gravity, by revolving the container.

A special development is the autogenous or semiautogenous mill. Autogenous mills operate without grinding bodies; instead, the coarser part of the ore simply grinds itself and the smaller fractions. To semiautogenous mills (which have become widespread), 5 to 10 percent grinding bodies (usually metal spheres) are added.

Roll Crusher

Yet another development, combining the processes of crushing and grinding, is the roll crusher. This consists essentially of two cylinders that are mounted on horizontal shafts and driven in opposite directions. The cylinders are pressed together under high pressure, so that comminution takes place in the material bed between them.

Concentration

Concentration involves the separation of valuable minerals from the other raw materials received from the grinding mill. In large-scale operations this is accomplished by taking advantage of the different properties of the minerals to be separated. These properties can be colour (optical sorting), density (gravity separation), magnetic or electric (magnetic and electrostatic separation), and physicochemical (flotation separation).

Optical Separation

This process is used for the concentration of particles that have sufficiently different colours (the best contrast being black and white) to be detected by the naked eye. In addition, electro-optic detectors collect data on the responses of minerals when exposed to infrared, visible, and ultraviolet light. The same principle, only using gamma radiation is called radiometric separation.

Gravity Separation

Gravity methods use the difference in the density of minerals as the concentrating agent.

In heavy-media separation (also called sink-and-float separation), the medium used is a suspension in water of a finely ground heavy mineral (such as magnetite or arsenopyrite) or technical product (such as ferrosilicon). Such a suspension can simulate a fluid with a higher density than water. When ground ores are fed into the suspension, the gangue particles, having a lower density, tend to float and are removed as tailings, whereas the particles of valuable minerals, having higher density, sink and are also removed. The magnetite or ferrosilicon can be removed from the tailings by magnetic separation and recycled.

In the process called jigging, a water stream is pulsed, or moved by pistons upward and downward, through the material bed. Under the influence of this oscillating motion, the bed is separated into layers of different densities, the heaviest concentrate forming the lowest layer and the lightest product the highest. Important to this process is a thorough classification of the feed, since particles less than one millimetre in size cannot be separated by jigging.

Finer-grained particles (from 1 millimetre to 50 micrometres) can be effectively separated in a flowing stream of water on horizontal or inclined planes. Most systems employ additional forces—for example, centrifugal force on spirals or impact forces on shaking tables. Spirals consist of a vertical spiral channel with an oval cross section. As the pulp flows from the top to the bottom of the channel, heavier particles concentrate on the inner side of the stream, where they can be removed through special openings. Owing to their low energy costs and simplicity of operation, the use of spirals has increased rapidly. They are especially effective at concentrating heavy mineral sands and gold ores.

Gravity concentration on inclined planes is carried out on shaking tables, which can be smoothed or grooved and which are vibrated back and forth at right angles to the flow of water. As the pulp flows down the incline, the ground material is stratified into heavy and light layers in the water; in addition, under the influence of the vibration, the particles are separated in the impact direction. Shaking tables are often used for concentrating finely grained ores of tin, tungsten, niobium, and tantalum.

Flotation Separation

Flotation is the most widely used method for the concentration of fine-grained minerals. It takes advantage of the different physicochemical surface properties of minerals—in particular, their wettability, which can be a natural property or one artificially changed by chemical reagents. By altering the hydrophobic (water-repelling) or hydrophilic (water-attracting) conditions of their surfaces, mineral particles suspended in water can be induced to adhere to air bubbles passing through a flotation cell or to remain in the pulp. The air bubbles pass to the upper surface of the pulp and form a froth, which, together with the attached hydrophobic minerals, can be removed. The tailings, containing the hydrophilic minerals, can be removed from the bottom of the cell.

Flotation makes possible the processing of complex intergrown ores containing copper, lead, zinc, and pyrite into separate concentrates and tailings—an impossible task with gravity, magnetic, or electric separation methods. In the past, these metals were recoverable only with expensive metallurgical processes.

Magnetic Separation

Magnetic separation is based on the differing degrees of attraction exerted on various minerals by magnetic fields. Success requires that the feed particles fall within a special size spectrum (0.1 to 1 millimetre). With good results, strongly magnetic minerals such as magnetite, franklinite, and pyrrhotite can be removed from gangue minerals by low-intensity magnetic separators. High-intensity devices can separate oxide iron ores such as limonite and siderite as well as iron-bearing manganese, titanium, and tungsten ores and iron-bearing silicates.

Electrostatic Separation

The electrostatic method separates particles of different electrical charges and, when possible, of different sizes. When particles of different polarity are brought into an electrical field, they follow different motion trajectories and can be caught separately. Electrostatic separation is used in all plants that process heavy mineral sands bearing zircon, rutile, and monazite. In addition, the cleaning of special iron ore and cassiterite concentrates as well as the separation of cassiterite-scheelite ores are conducted by electrostatic methods.

Dewatering

Concentrates and tailings produced by the methods outlined above must be dewatered in order to convert the pulps to a transportable state. In addition, the water can be recycled into the existing water circuits of the processing plant, greatly reducing the demand for expensive fresh water.

Filtration

Filtration is the separation of a suspension into a solid filter cake and a liquid filtrate by passing it through a permeable filtering material. Important factors in this process are the properties of the suspension (e.g., size distribution, concentration), the properties of the filtering materials (e.g., the width and shape of pores), and the forces applied to the suspension. Filtration is carried out in gravity filters (screens, dewatering bins), in centrifugal filters (screen centrifuges), in vacuum filters (drum cell filters, disk filters), or in pressure filters (filter presses). Such devices make it possible to produce filter cakes containing 8 to 15 percent moisture.

Thickening

In the process of thickening (also called sedimentation), the solids in a suspension settle under the influence of gravity in a tank and form a thick pulp. This pulp, and the clear liquid at the top of the tank, can be removed continuously or intermittently. In comparison with filtration, thickening offers the advantage of low operation costs; on the other hand, it has the disadvantage of leaving a higher moisture content in the pulp. For this reason, the dewatering of pulps containing fine particles often involves a combination of thickening and filtration. The thickening of finely grained pulps is often aided by the use of flocculating agents.

Drying

The removal of water from solid materials by thermal drying plays a significant role in modern mineral processing. A great number of dryer types are available. Convection dryers, employing a flow of hot combustion gases to remove moisture from a pulp stream, are the most common. To this type belong rotary drum, conveyor, and fluidized-bed dryers.

SMELTING

Electric phosphate smelting furnace in a TVA chemical plant.

Smelting is a process of applying heat to ore in order to extract a base metal. It is a form of extractive metallurgy. It is used to extract many metals from their ores, including silver, iron, copper, and other base metals. Smelting uses heat and a chemical reducing agent to decompose the ore, driving off other elements as gases or slag and leaving the metal base behind. The reducing agent is commonly a source of carbon, such as coke—or, in earlier times, charcoal.

The carbon (or carbon monoxide derived from it) removes oxygen from the ore, leaving the elemental metal. The carbon thus oxidizes in two stages, producing first carbon monoxide and then carbon dioxide. As most ores are impure, it is often necessary to use flux, such as limestone, to remove the accompanying rock gangue as slag.

Plants for the electrolytic reduction of aluminium are also generally referred to as aluminium smelters.

Labourers working in the smelting industry have reported respiratory illnesses inhibiting their ability to perform the physical tasks demanded by their jobs.

Process

Smelting involves more than just melting the metal out of its ore. Most ores are the chemical compound of the metal and other elements, such as oxygen (as an oxide), sulfur (as a sulfide), or carbon and oxygen together (as a carbonate). To extract the metal, workers must make these compounds undergo a chemical reaction. Smelting therefore consists of using suitable reducing substances that combine with those oxidizing elements to free the metal.

Roasting

In the case of carbonates and sulfides, a process called "roasting" drives out the unwanted carbon or sulfur, leaving an oxide, which can be directly reduced. Roasting is usually carried out in an oxidizing environment. A few practical examples:

- Malachite, a common ore of copper, is primarily copper carbonate hydroxide $Cu_2(CO_3)$ $(OH)_2$. This mineral undergoes thermal decomposition to $2CuO$, CO_2, and H_2O in several stages between 250 °C and 350 °C. The carbon dioxide and water are expelled into the atmosphere, leaving copper(II) oxide, which can be directly reduced to copper.

- Galena, the most common mineral of lead, is primarily lead sulfide (PbS). The sulfide is oxidized to a sulfite ($PbSO_3$), which thermally decomposes into lead oxide and sulfur dioxide gas. (PbO and SO_2) The sulfur dioxide is expelled and the lead oxide is reduced.

Reduction

Reduction is the final, high-temperature step in smelting, in which the oxide becomes the elemental metal. A reducing environment (often provided by carbon monoxide, made by incomplete combustion in an air-starved furnace) pulls the final oxygen atoms from the raw metal. The required temperature varies over a very large range, both in absolute terms and in terms of the melting point of the base metal. Examples:

- Iron oxide becomes metallic iron at roughly 1250 °C (2282 °F or 1523.15 K), almost 300 degrees *below* iron's melting point of 1538 °C (2800.4 °F or 1811.15 K).

- Mercuric oxide becomes vaporous mercury near 550 °C (1022 °F or 823.15 K), almost 600 degrees *above* mercury's melting point of -38 °C (-36.4 °F or 235.15 K).

Flux and slag can provide a secondary service after the reduction step is complete: they provide a molten cover on the purified metal, preventing contact with oxygen while still hot enough to readily oxidize. This prevents impurities from forming in the metal.

Fluxes

Metal workers use fluxes in smelting for several purposes, chief among them catalyzing the desired reactions and chemically binding to unwanted impurities or reaction products. Calcium oxide, in the form of lime, was often used for this purpose, since it could react with the carbon dioxide and sulfur dioxide produced during roasting and smelting to keep them out of the working environment.

Base Metals

Cowles Syndicate of Ohio in Stoke-upon-Trent England, late 1880s. British
Aluminium used the process of Paul Héroult about this time.

The ores of base metals are often sulfides. In recent centuries, reverberatory furnaces have
been used to keep the charge being smelted separate from the fuel. Traditionally, they were
used for the first step of smelting: forming two liquids, one an oxide slag containing most of
the impurities, and the other a sulfide matte containing the valuable metal sulfide and some
impurities. Such "reverb" furnaces are today about 40 meters long, 3 meters high and 10 meters wide. Fuel is burned at one end to melt the dry sulfide concentrates (usually after partial
roasting) which are fed through openings in the roof of the furnace. The slag floats over the
heavier matte and is removed and discarded or recycled. The sulfide matte is then sent to the
converter. The precise details of the process vary from one furnace to another depending on the
mineralogy of the orebody.

While reverberatory furnaces produced slags containing very little copper, they were relatively
energy inefficient and off-gassed a low concentration of sulfur dioxide that was difficult to capture; a new generation of copper smelting technologies has supplanted them. More recent furnaces exploit bath smelting, top-jetting lance smelting, flash smelting and blast furnaces. Some
examples of bath smelters include the Noranda furnace, the Isasmelt furnace, the Teniente reactor, the Vunyukov smelter and the SKS technology. Top-jetting lance smelters include the
Mitsubishi smelting reactor. Flash smelters account for over 50% of the world's copper smelters.
There are many more varieties of smelting processes, including the Kivset, Ausmelt, Tamano,
EAF, and BF.

Environmental Implications

Smelting seriously impacts the environment by producing wastewater and slag and releasing such
toxic metals as copper, silver, iron, cobalt and selenium into the atmosphere. It also releases gaseous sulfur dioxide, contributing to acid rain, which acidifies soil and water.

HYDROMETALLURGY

Hydrometallurgy is the extraction of metal from ore by preparing an aqueous solution of a salt of the metal and recovering the metal from the solution. The operations usually involved are leaching, or dissolution of the metal or metal compound in water, commonly with additional agents; separation of the waste and purification of the leach solution; and the precipitation of the metal or one of its pure compounds from the leach solution by chemical or electrolytic means. The most common leaching agent is dilute sulfuric acid.

Hydrometallurgy originated in the 16th century, but its principal development took place in the 20th century, stimulated partly by the desire to extract gold from low-grade ores. The development of ion exchange, solvent extraction, and other processes has led to an extremely broad range of applications of hydrometallurgy, now used to produce more than 70 metallic elements. Besides most gold and much silver, large tonnages of copper and zinc are produced by hydrometallurgy.

Hydrometallurgy of Copper

The most common application of hydrometallurgy to copper ores is to leach the low-grade and oxidized rock that surrounded the higher-grade copper sulfide deposits. With increasingly stringent environmental regulations, it is becoming more difficult to get the permits for a copper smelter, and so there is interest in developing hydrometallurgical processes that can completely replace conventional smelters. At the moment, hydrometallurgy is not competitive with pyrometallurgy for producing metal from high-grade ores.

Ore Preparation

The ore will require some degree of comminution, so that the leaching solutions will be able to reach the copper oxides and sulfides that are to be dissolved, as shown in figure. The amount of comminution is limited by the value of the ore. For example, a very low-grade copper ore might only be crushed to a two inch top size, because the value of the added copper that would be recovered by grinding finer is not enough to justify the cost.

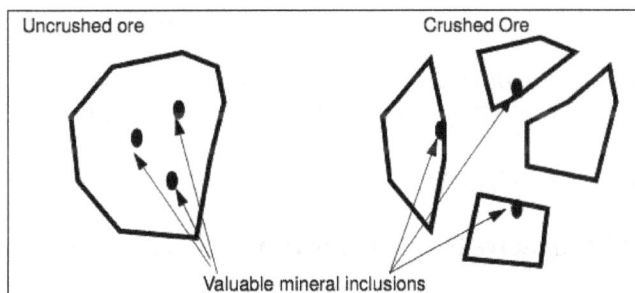

In uncrushed ore, valuable mineral inclusions are completely surrounded by rock,
and cannot be reached easily by leaching solutions. Crushing the ore exposes these inclusions to leaching.

When possible, it is helpful to pre-concentrate the ore, to reject any rock that contains very little copper. This reduces the amount of material that has to be leached, and so reduces the size of the plant necessary to deal with it.

Leaching

Copper leaching is the dissolution of the copper in a solvent, while leaving the gangue minerals behind as undissolved solids. Copper is normally leached using one of the following methods:

- Dump leaching

- Heap leaching

Dump Leaching: The overburden and tailings dumps for copper mining operations often contain enough copper that it would be worthwhile to recover some of it. Dump leaching is then used to recover as much of the copper as can be leached out without spending a lot of time and money on preparing the ore. This is done by trickling the leaching solution over a dump, and collecting the runoff solution to recover the copper that it dissolves. Dump leaching is quite slow, with periods of months or years needed before leaching is completed, and typically only about 60% of the copper in the dump is recovered.

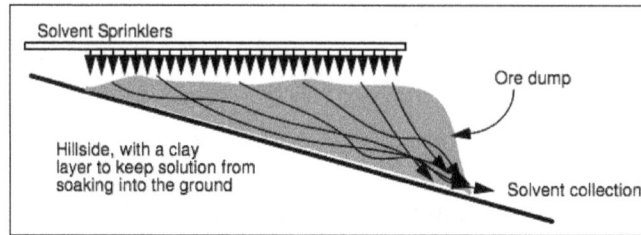

Typical dump-leaching operation.

Heap Leaching: This is similar to dump leaching, but instead of simply dumping ore on a hillside, the ore is crushed approximately to the size of gravel (to improve leaching rates and recovery) and piled onto artificial waterproof pads, as shown in figure. Once the ore has finished leaching (after approximately 6 months to a year), the leached gangue is removed from the pad for disposal, and replaced with fresh ore.

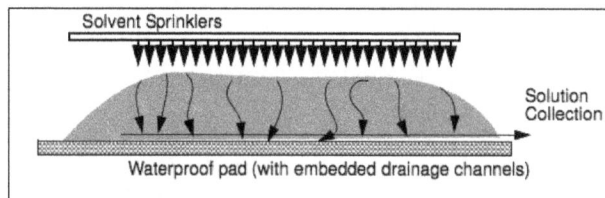

Typical heap-leaching operation. The pad is usually concrete, with a rubber or asphalt coating to make it waterproof.

Factors affecting the Leaching Rate

The amount of time needed to dissolve the copper from the ore is influenced by the following factors:

- Particle Size: Larger particles take longer to leach completely than smaller particles. The reason for this can be seen from the shrinking-core model of leaching. When a particle first begins to leach, the leaching solution can diffuse quickly to the portion of the ore that contains metal, and then diffuse back out. As time passes, the core of unleached ore becomes smaller, and the shell of leached ore (which the solution must diffuse through to reach the core) becomes thicker. Since diffusion is a slow process, this means that the leaching of a particle goes more slowly as time goes by. Also, since larger particles will have a thicker shell of leached material, they will take much longer to leach completely that small particles will.

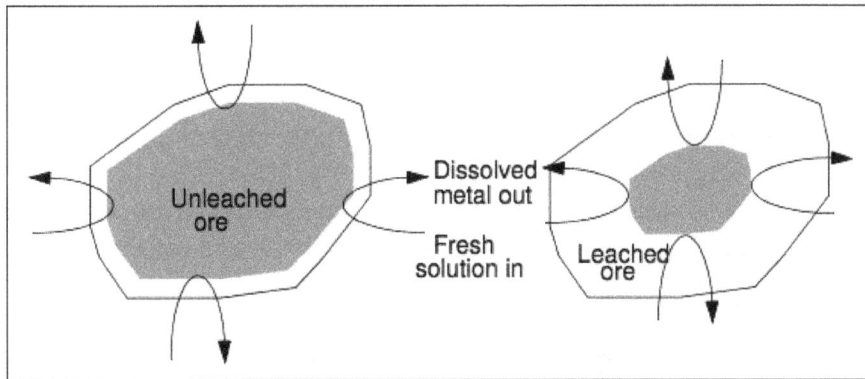

The Shrinking-Core model. As the shell of leached ore
becomes thicker, it becomes more difficult for solvent to reach
the unleached ore, and so the leaching rate slows down.

- Reagent Concentration: The speed of the dissolution can often be increased by adding more of the leaching reagents (acids, complexing agents, oxidizers, etc.). At very high reagent concentrations, the cost will become prohibitive.

- Temperature: Changing the temperature will increase the rate of metal dissolution, increase the rate of diffusion, and usually increase the total amount of metal that can be dissolved per liter of solvent. Increases in temperature are limited by the boiling point of the leaching solution.

- Thoroughness of Mixing: Improving the mixing between the ore and the solvent will reduce the number of "dead zones", which leach slowly because they do not contact fresh solvent. By mixing the ore and the solvent thoroughly, the speed and completeness of leaching will increase. Eventually, the ore and solvent will be perfectly mixed, and further improvements will not be possible.

Types of Leaching Reactions

Depending on the nature of the ore, copper leaching can either be a non-oxidative process, or an oxidative process.

Non-oxidative leaching: In non-oxidative leaching, there is no change in oxidation state. One example is the dissolution of copper sulfate by water:

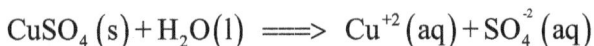

$$CuSO_4(s) + H_2O(l) \implies Cu^{+2}(aq) + SO_4^{-2}(aq)$$

Another type of non-oxidative leaching reaction is the dissolution of alkaline materials, such as malachite, by an acid:

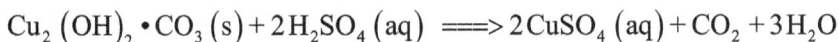

$$Cu_2(OH)_2 \cdot CO_3(s) + 2H_2SO_4(aq) \implies 2CuSO_4(aq) + CO_2 + 3H_2O$$

Oxidative leaching: Many ores are not soluble unless they are oxidized. For example, CuS (covellite) is made much more soluble if it is oxidized to copper sulfate:

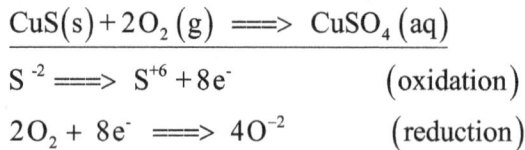

$$CuS(s) + 2O_2(g) ===> CuSO_4(aq)$$
$$S^{-2} ===> S^{+6} + 8e^- \qquad (oxidation)$$
$$2O_2 + 8e^- ===> 4O^{-2} \qquad (reduction)$$

In this case, it is the oxidation of sulfur to the more soluble SO_4^{-2} that makes the material soluble. The oxidative leaching reaction above uses oxygen as the oxidizing agent. Unfortunately, gaseous oxygen is not very soluble in water (at 20°C, water in equilibrium with air only dissolves about 8 parts per million oxygen). Because of this, it is difficult to dissolve enough oxygen in the leaching solution to give a rapid leaching rate, and violent agitation of the slurry is needed to bring air in contact with the solid particles that need to be oxidized. A solution to this is to use other oxidizing agents, that are more soluble in water than oxygen is. One such oxidizing agent is ferric iron (Fe^{+3}).

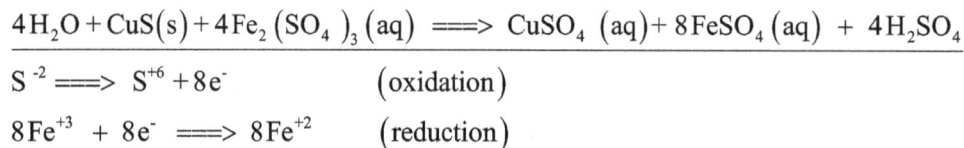

$$4H_2O + CuS(s) + 4Fe_2(SO_4)_3(aq) ===> CuSO_4(aq) + 8FeSO_4(aq) + 4H_2SO_4$$
$$S^{-2} ===> S^{+6} + 8e^- \qquad (oxidation)$$
$$8Fe^{+3} + 8e^- ===> 8Fe^{+2} \qquad (reduction)$$

There are two drawbacks to using ferric iron as an oxidizing agent. The first is that it is only soluble in acidic solutions (pH less than 4). If the solution approaches neutral pH, then the ferric iron will rapidly precipitate out as iron hydroxide, and be useless. The second problem is that, unlike air, it is not free, and so to be economical it should be regenerated by oxidizing the ferrous iron (Fe^{+2}) with air to convert it back to ferric iron.

Bacterial leaching: The re-oxidation of ferrous iron to ferric iron would be too slow to be of practical use, except for the action of the bacteria Thiobacillus ferrooxidans. This bacteria is unusual, in that it gets its metabolic energy by oxidizing ferrous iron to ferric iron, rather than by eating organic material or by collecting sunlight like other organisms do. In nature, this bacteria and its relatives oxidize pyrite (FeS_2) to produce dissolved ferric sulfate and sulfuric acid, according to the reactions:

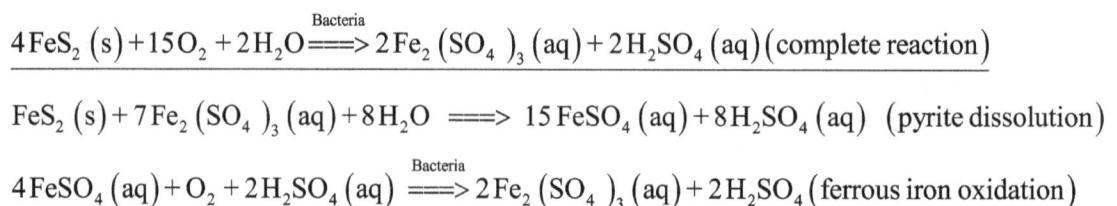

$$4FeS_2(s) + 15O_2 + 2H_2O \overset{Bacteria}{===>} 2Fe_2(SO_4)_3(aq) + 2H_2SO_4(aq) \ (complete \ reaction)$$

$$FeS_2(s) + 7Fe_2(SO_4)_3(aq) + 8H_2O ===> 15FeSO_4(aq) + 8H_2SO_4(aq) \ (pyrite \ dissolution)$$

$$4FeSO_4(aq) + O_2 + 2H_2SO_4(aq) \overset{Bacteria}{===>} 2Fe_2(SO_4)_3(aq) + 2H_2SO_4 \ (ferrous \ iron \ oxidation)$$

Both the ferric iron and the sulfuric acid are useful leaching reagents for copper sulfides, and pyrite is a common accessory mineral in copper sulfide mineral deposits. As a result, these bacteria can greatly reduce the cost of leaching operations by producing many of the necessary reagents directly from the ore. Thiobacillus ferrooxidans and related species are widespread, and applications such as heap and dump leaching (which are exposed to the open air) will rapidly pick up an appropriate bacterial culture without being inoculated.

Solution Purification

Leaching reactions are not perfectly selective, and so there will be many elements in solution after leaching, and not just copper. These other elements can cause problems in recovering the metal, and so they sometimes need to be removed from the solution. Also, the copper can be very dilute, and so a means is needed to concentrate it. This is generally done using ion exchange processes.

Ion-exchange Groups

Ion-exchange separations are based on the electrostatic attraction of certain molecular groups for ions in solution. These groups can be either positively charged or negatively charged, as shown in figure. The groups are attached to any of several types of larger molecular structures, so that they will not go into solution and be lost. The groups are commonly attached to either solid resins, or to organic liquids that are not soluble in water.

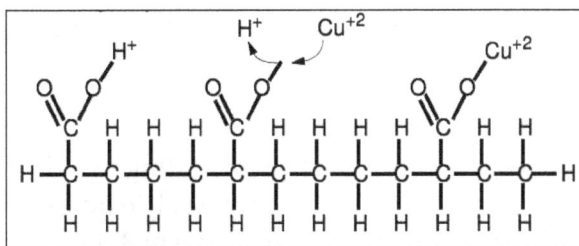

The above figure shows, mechanism of ion exchange in solvent extraction. Active groups such as carboxyl are attached to the hydrocarbon chain of a non-water-soluble oil. When the oil is contacted with water, the carboxyl exchanges the ion it currently holds (H^+ in this case) for a copper ion. It can be made to release the attached copper by contacting it with a stripping solution that has a very high H^+ concentration.

Solvent Extraction

In solvent extraction, the ion-exchange groups are attached to the molecules of an organic liquid that is insoluble in water. This makes it possible to make a continuous process.

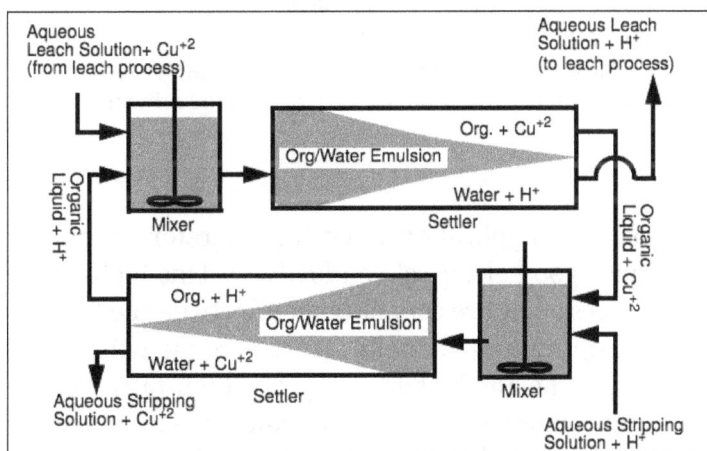

Solvent-Extraction process. The "mixers" ensure good mixing between the water solutions and the organic liquid so that ions can be freely passed back and forth. The "settlers" give the organic liquid the chance to separate from the water.

Metal Recovery

Once the metal has been dissolved, the solids have been removed, and the solution has been purified, the metal must be recovered in a solid form. This can be done chemically, or electrochemically.

Cementation

Dissolved copper will plate out on an iron surface due to the following reaction:

$$Cu^{+2}(aq) + Fe(s) \implies Fe^{+2}(aq) + Cu(s)$$

So, if a solution with dissolved copper is run through a bed of shredded scrap iron, the iron will oxidize and dissolve, while reducing the copper ions and plating them out as solid copper.

For cementation to be efficient, the iron scrap must have a high surface area.

Electrowinning

Electrowinning is an electrochemical process for precipitating metals from solution, as shown in figure. The anode is made out of a material that will not easily oxidize and dissolve, such as lead or titanium. The advantage of electrowinning is that once the metal is plated out of the solution, the metal-depleted solution can usually be recycled directly to the hydrometallurgy operation.

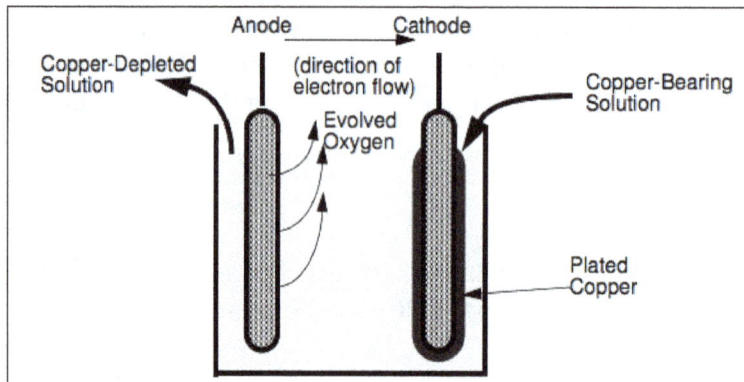

Basic electrowinning cell. Oxygen gas is produced
at the anode, while metal is plated onto the cathode.

Electrorefining

Most copper is used for electrical applications, and it is therefore important that the copper be high-purity so that it will have good electrical conductivity. This high-purity copper is produced by electrorefining. In this process, impure copper (generally from the fire-refining stage in a pyrometallurgical operation) is cast into copper anodes. These are then placed into an electrochemical cell which dissolves metal from the anode, and redeposits high-purity copper on the cathode, as shown in figure. Metals that are more electropositive than copper (such as zinc and iron) dissolve and remain in solution, while metals that are less electropositive than copper (such as gold and silver) accumulate as an "anode sludge" that is collected and sold for further refining.

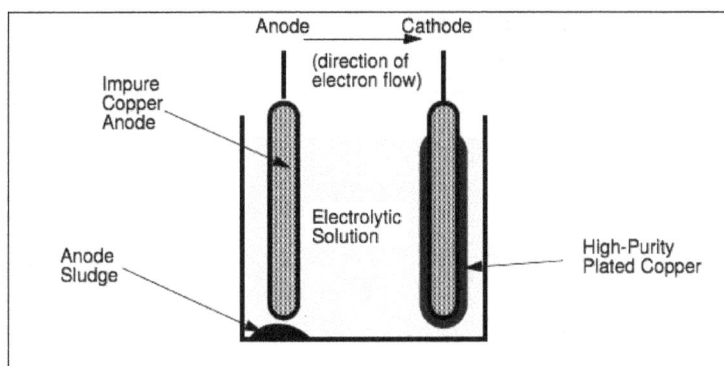

Basic electrorefining cell for producing high-purity copper.

Layout of a Typical Hydrometallurgical Operation for Copper

A copper dump-leaching operation would generally be laid out roughly as shown in figure. Sulfuric acid would normally be produced by the associated copper smelter, and other reagents are recycled as much as possible both to minimize effluents and to reduce reagent costs.

Since organic liquid from the solvent extraction will cause problems if it gets into the electrowinning cells, a froth flotation stage is often included between solvent extraction and electrowinning to completely remove any remaining droplets of organic liquid.

For a simpler circuit, the solvent extraction/electrowinning stages could be replaced by a cementation step, which would produce a lower-purity copper product that would need to be refined further.

Leaching

Leaching is a widely used extractive metallurgy technique which converts metals into soluble salts in aqueous media. Compared to pyrometallurgical operations, leaching is easier to perform and much less harmful, because no gaseous pollution occurs. The only drawback of leaching is its lower efficiency caused by the low temperatures of the operation, which dramatically affect chemical reaction rates.

There are a variety of leaching processes, usually classified by the types of reagents used in the operation. The reagents required depend on the ores or pretreated material to be processed. A typical feed for leaching is either oxide or sulfide.

For material in oxide form, a simple acid leaching reaction can be illustrated by the zinc oxide leaching reaction:

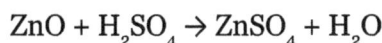

$$ZnO + H_2SO_4 \rightarrow ZnSO_4 + H_2O$$

In this reaction solid ZnO dissolves, forming soluble zinc sulfate.

In many cases other reagents are used to leach oxides. For example, in the metallurgy of aluminium, aluminium oxide is subject to leaching by alkali solutions:

$$Al_2O_3 + 3H_2O + 2NaOH \rightarrow 2NaAl(OH)_4$$

Leaching of sulfides is a more complex process due to the refractory nature of sulfide ores. It often involves the use of pressurized vessels, called autoclaves. A good example of the autoclave leach process can be found in the metallurgy of zinc. It is best described by the following chemical reaction:

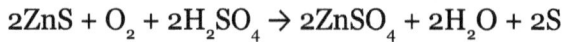

$$2ZnS + O_2 + 2H_2SO_4 \rightarrow 2ZnSO_4 + 2H_2O + 2S$$

This reaction proceeds at temperatures above the boiling point of water, thus creating a vapor pressure inside the vessel. Oxygen is injected under pressure, making the total pressure in the autoclave more than 0.6 MPa.

The leaching of precious metals such as gold can be carried out with cyanide or ozone under mild conditions.

Types of Leaching

Vat Leaching

Vat Leaching is also known as "agitated tank" leaching. In this the lixiviant comes into contact with the metal material in large vats or tanks, which can be stirred, enhancing the reaction kinetics. The metal containing solid, such as concentrate, ore, residue, or slag, often undergoes size reduction via crushing and grinding prior to leaching.

One example of a leaching process that uses vat leaching is gold cyanidation, the process of extracting gold from low-grade ores. In this process, a dilute solution of sodium cyanide (NaCN) is used to leach the Au into solution. The concentration is usually 0.01-0.05 % cyanide, or 100-500ppm. It should also be noted that the alkalinity of this solution must be high enough such that hydrogen cyanide is not produced, which is a very toxic gas. This is more important when sulfide minerals are in the ore, because they will oxidize, consuming oxygen and generating acid. The pH can be kept above 10, by adding a base such as lime (CaO) to the solution, at a minimum concentration of 0.04%, when NaCN is 0.01%. Other additives include lead nitrate solution (at a ratio of 2:3 to the NaCN solution) in order to minimize cyanide consumption in the reaction. Based on different variations of gold cyanide leaching and the starting material, different concentrations and ratios of the solvents and additives are used.

Tank Leaching

In metallurgical processes tank leaching is a hydrometallurgical method of extracting valuable material (usually metals) from ore.

Tank vs. Vat Leaching

Factors

Tank leaching is usually differentiated from vat leaching on the following factors:

1. In tank leaching the material is ground sufficiently fine to form a slurry or pulp, which can flow under gravity or when pumped. In vat leaching typically a coarser material is placed in the vat for leaching, this reduces the cost of size reduction.

2. Tanks are typically equipped with agitators, baffles, gas introduction equipment designed to maintain the solids in suspension in the slurry, and achieve leaching. Vats usually do not contain much internal equipment, except for agitators.

3. Tank leaching is typically continuous, while vat leaching is operated in a batch fashion, this is not always the case, and commercial processes using continuous vat leaching have been tested.

4. Typically the retention time required for vat leaching is more than that for tank leaching to achieve the same percentage of recovery of the valuable material being leached.

In a tank leach the slurry is moved, while in a vat leach the solids remain in the vat, and solution is moved.

Processes

Tank and vat leaching involves placing ore, usually after size reduction and classification, into large tanks or vats at ambient operating conditions containing a leaching solution and allowing the valuable material to leach from the ore into solution.

In tank leaching the ground, classified solids are already mixed with water to form a slurry or pulp, and this is pumped into the tanks. Leaching reagents are added to the tanks to achieve the leaching reaction. In a continuous system the slurry will then either overflow from one tank to the next, or be pumped to the next tank. Ultimately the "pregnant" solution is separated from the slurry using some form of liquid/solid separation process, and the solution passes on to the next phase of recovery.

In vat leaching the solids are loaded into the vat, once full the vat is flooded with a leaching solution. The solution drains from the tank, and is either recycled back into the vat or is pumped to the next step of the recovery process. . Vat leach units are rectangular containers (drums, barrels, tanks or vats), usually very big and made of wood or concrete, lined with material resistant to the leaching media. The treated ore is usually coarse.

The vats are usually run sequentially to maximize the contact time between the ore and the reagent. In such a series the leachate collected from one container is added to another vat with fresher ore.

The tanks are equipped with agitators to keep the solids in suspension in the vats and improve the solid to liquid to gas contact. Agitation is further assisted by the use of tank baffles to increase the efficiency of agitation and prevent centrifuging of slurries in circular tanks.

Extraction Efficiency Factors

Aside from chemical requirements several key factors influence extraction efficiency:

- Retention time: Refers to the time spent in the leaching system by the solids. This is calculated as the total volumetric capacity of the leach tank/s divided by the volumetric throughput of the solid/liquid slurry. Retention time is commonly measured in hours for precious metals recovery. A sequence of leach tanks is referred to as a leach "train", and retention time is measured considering the total volume of the leach train. The desired retention time is determined during the testing phase, and the system is then designed to achieve this.

- Size: The ore must be ground to a size that exposes the desired mineral to the leaching agent (referred to as "liberation"), and in tank leaching this must be a size that can be suspended by the agitator. In vat leaching this is the size that is the most economically viable, where the recovery achieved as ore is ground finer is balanced against the increased cost of processing the material.

- Slurry density: The slurry density (percent solids) determines retention time. The settling rate and viscosity of the slurry are functions of the slurry density. The viscosity, in turn, controls the gas mass transfer and the leaching rate.

- Numbers of tanks: Agitated tank leach circuits are typically designed with no less than four tanks and preferably more to prevent short-circuiting of the slurry through the tanks.

- Dissolved gas: Gas is often injected below the agitator or into the vat to obtain the desired dissolved gas levels – typically oxygen, in some base metal plants sulphur dioxide may be required.

- Reagents: Adding and maintaining the appropriate amount of reagents throughout the leach circuit is critical to a successful operation. Adding insufficient quantities of reagents reduces the metal recovery but adding excess reagents increases the operating costs without recovering enough additional metal to cover the cost of the reagents.

The tank leaching method is commonly used to extract gold and silver from ore, such as with the Sepro Leach Reactor.

Heap Leaching

Gold heap leaching.

Heap leaching is an industrial mining process used to extract precious metals, copper, uranium, and other compounds from ore using a series of chemical reactions that absorb specific minerals and re-separate them after their division from other earth materials. Similar to *in situ* mining, heap leach mining differs in that it places ore on a liner, then adds the chemicals via drip systems to the ore, whereas *in situ* mining lacks these liners and pulls pregnant solution up to obtain the minerals. Most mining companies favor the economic feasibility of heap leaching, considering that heap leaching is a better alternative to conventional processing methods such as such as flotation, agitation, and vat leaching.

Additionally, dump leaching is an essential part of most copper mining operations and determines the quality grade of the produced material along with other factors. Due to the profitability that the dump leaching has on the mining process, i.e. it can contribute substantially to the economic viability of the mining process, it is advantageous to include the results of the leaching operation in the economic overall project evaluation. This, in effect, requires that the key controllable variables, which have an effect on the recovery of the metal and the quality of solution coming from a dump leaching process.

The process has ancient origins; one of the classical methods for the manufacture of copperas (iron sulfate) was to heap up iron pyrite and collect the leachate from the heap, which was then boiled with iron to produce iron(II) sulfate.

Process

The mined ore is usually crushed into small chunks and heaped on an impermeable plastic or clay lined leach pad where it can be irrigated with a leach solution to dissolve the valuable metals. While sprinklers are occasionally used for irrigation, more often operations use drip irrigation to minimize evaporation, provide more uniform distribution of the leach solution, and avoid damaging the exposed mineral. The solution then percolates through the heap and leaches both the target and other minerals. This process, called the "leach cycle", generally takes from one or two months for simple oxide ores (e.g. most gold ores) to two years for nickel laterite ores. The leach solution containing the dissolved minerals is then collected, treated in a process plant to recover the target mineral and in some cases precipitate other minerals, and recycled to the heap after reagent levels are adjusted. Ultimate recovery of the target mineral can range from 30% of contained run-of-mine dump leaching sulfide copper ores to over 90% for the ores that are easiest to leach, some oxide gold ores.

Left: ore fines without agglomeration. Right: Ore fines after agglomeration - Improved percolation as a result of agglomeration.

The essential questions to address during the process of the heap leaching are as following:

1. Can the investment of crushing the ore be justified by the potential increase in recovery and rate of recovery?

2. How should the concentration of acid be altered over time in order to produce a solution that can be economically treated?

3. How does the form of a heap affect the recovery and solution grade?

4. Under any given set of circumstances, what type of recovery can be expected before the leach solution quality drops below a critical limit?

5. What recovery (quantifiable measure) can be expected?

In recent years, the addition of an agglomeration drum has improved on the heap leaching process by allowing for a more efficient leach. The rotary drum agglomerator, such as the tyre driven Sepro Agglomeration Drum works by taking the crushed ore fines and agglomerating them into more uniform particles. This makes it much easier for the leaching solution to percolate through the pile, making its way through the channels between particles.

The addition of an agglomeration drum also has the added benefit of being able to pre-mix the leaching solution with the ore fines to achieve a more concentrated, homogeneous mixture and allow the leach to begin prior to the heap.

Although heap leach design has made significant progress over the last few years through the use of new materials and improved analytical tools, industrial experience shows that there are significant benefits from extending the design process beyond the liner and into the rock pile itself. Characterization of the physical and hydraulic (hydrodynamic) properties of ore-for-leach focuses on the direct measurement of the key properties of the ore, namely:

- The relationship between heap height and ore bulk density (density profile).

- The relationship between bulk density and percolation capacity (conductivity profile).

- The relationship between the bulk density, porosity and its components (micro and macro).

- The relationship between the moisture content and percolation capacity (conductivity curve).

- The relationship between the aforementioned parameters and the ore preparation practices (mining, crushing, agglomeration, curing, and method of placement).

Theoretical and numerical analysis, and operational data show that these fundamental mechanisms are controlled by scale, dimensionality, and heterogeneity, all of which adversely affect the scalability of metallurgical and hydrodynamic properties from the lab to the field. The dismissal of these mechanisms can result in a number of practical and financial problems that will resonate throughout the life of the heap impacting the financial return of the operation. Through procedures that go beyond the commonly employed metallurgical testing and the integration of data gleaned through real time 3D monitoring, a more complete representative characterization of the physicochemical properties of the heap environment is obtained. This improved understanding results in a significantly higher degree of accuracy in terms of creating a truly representative sample of the environment within the heap.

By adhering to the characterization identified above, a more comprehensive view of heap leach environments can be realized, allowing the industry to move away from the *de facto* black-box approach to a physicochemically inclusive industrial reactor model.

Precious Metals

The crushed ore is irrigated with a dilute alkaline cyanide solution. The solution containing the dissolved precious metals in a pregnant solution continues percolating through the crushed ore until it reaches the liner at the bottom of the heap where it drains into a storage (pregnant solution) pond. After separating the precious metals from the pregnant solution, the dilute cyanide solution (now called "barren solution") is normally re-used in the heap-leach-process or occasionally sent to an industrial water treatment facility where the residual cyanide is treated and residual metals are removed. In very high rainfall areas, such as the tropics, in some cases there is surplus water that is then discharged to the environment, after treatment, posing possible water pollution if treatment is not properly carried out.

The production of one gold ring through this method, can generate 20 tons of waste material.

During the extraction phase, the gold ions form complex ions with the cyanide:

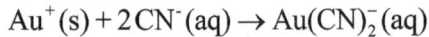

$$Au^+(s) + 2CN^-(aq) \rightarrow Au(CN)_2^-(aq)$$

Recuperation of the gold is readily achieved with a redox-reaction:

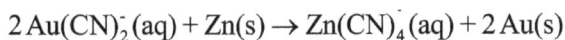

$$2Au(CN)_2^-(aq) + Zn(s) \rightarrow Zn(CN)_4^-(aq) + 2Au(s)$$

The most common methods to remove the gold from solution are either using activated carbon to selectively absorb it, or the Merrill-Crowe process where zinc powder is added to cause a precipitation of gold and zinc. The fine product can be either doré (gold-silver bars) or zinc-gold sludge that is then refined elsewhere.

Copper Ores

The method is similar to the cyanide method above, except sulfuric acid is used to dissolve copper from its ores. The acid is recycled from the solvent extraction circuit (see solvent extraction-electrowinning, SX/EW) and reused on the leach pad. A byproduct is iron(II) sulfate, jarosite, which is produced as a byproduct of leaching pyrite, and sometimes even the same sulfuric acid that is needed for the process. Both oxide and sulfide ores can be leached, though the leach cycles are much different and sulfide leaching requires a bacterial, or bio-leach, component.

In 2011 leaching, both heap leaching and in-situ leaching, produced 3.4 million metric tons of copper, 22 percent of world production. The largest copper heap leach operations are in Chile, Peru, and the southwestern United States.

Although heap leaching is a low cost-process, it normally has recovery rates of 60-70%. It is normally most profitable with low-grade ores. Higher-grade ores are usually put through more complex milling processes where higher recoveries justify the extra cost. The process chosen depends on the properties of the ore.

The final product is cathode copper.

Nickel Ores

This method is an acid heap leaching method like that of the copper method in that it utilises sulfuric acid instead of cyanide solution to dissolve the target minerals from crushed ore. The amount of sulfuric acid required is much higher than for copper ores, as high as 1,000 kg of acid per tonne of ore, but 500 kg is more common. The method was originally patented by Australian miner BHP Billiton and is being commercialized by Cerro Matoso S.A. in Colombia, a wholly owned subsidiary of BHP Billiton; Vale in Brazil; and European Nickel PLC for the rock laterite deposits of Turkey, Talvivaara mine in Finland, the Balkans, and the Philippines. There currently are no operating commercial scale nickel laterite heap leach operations, but there is a sulphide HL operating in Finland.

Nickel recovery from the leach solutions is much more complex than for copper and requires various stages of iron and magnesium removal, and the process produces both leached ore residue ("ripios") and chemical precipitates from the recovery plant (principally iron oxide residues, magnesium sulfate and calcium sulfate) in roughly equal proportions. Thus, a unique feature of nickel heap leaching is the need for a tailings disposal area.

The final product can be nickel hydroxide precipitates (NHP) or mixed metal hydroxide precipitates (MHP), which are then subject to conventional smelting to produce metallic nickel.

Uranium Ores

The final product is yellowcake and requires significant further processing to produce fuel-grade feed.

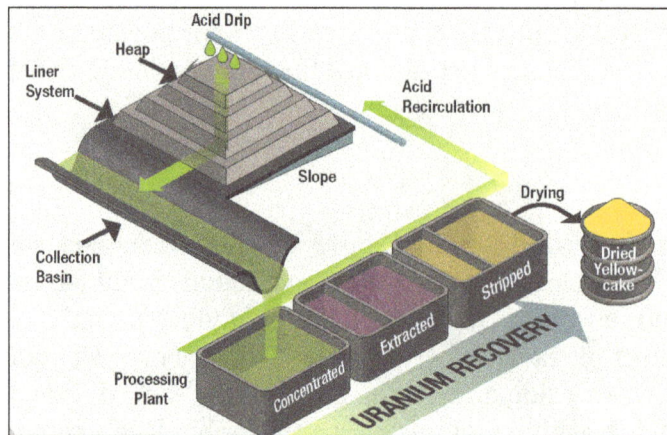

Diagram of heap leach recovery for uranium (US NRC).

Similar to copper oxide heap leaching, also using dilute sulfuric acid. Rio Tinto is commercializing this technology in Namibia and Australia; the French nuclear power company Areva, in Niger with two mines and Namibia; and several other companies are studying its feasibility.

Apparatus

While most mining companies have shifted from a previously accepted sprinkler method to the percolation of slowly dripping choice chemicals including cyanide or sulfuric acid closer to the actual ore bed, heap leach pads have not changed too much throughout the years. There are still four main categories of pads: conventional, dump leach, Valley Fills, and on/off pads. Typically, each pad only has a single, geomembrane liner for each pad, with a minimum thickness of 1.5mm, usually thicker.

The conventional pads simplest in design are used for mostly flat or gentle areas and hold thinner layers of crushed ore. Dump leach pads hold more ore and can usually handle a less flat terrain. Valley Fills are pads situated at valley bottoms or levels that can hold everything falling into it. On/off pads involve the use of putting significantly larger loads on the pads and removing and reloading it after every cycle.

Many of these mines which previously had digging depths of about 15 meters are digging deeper than ever before to mine materials, approximately 50 meters, sometimes more, which means that, in order to accommodate all of the ground being displaced, pads will have to hold higher weights from more crushed ore being contained in a smaller area. With that increase in build up comes in potential for decrease in yield or ore quality, as well as potential either weak spots in the lining or areas of increased pressure buildup. This build up still has the potential to lead to punctures in the liner. As of 2004 cushion fabrics, which could reduce potential punctures and their leaking, were still being debated due to their tendency to increase risks if too much weight on too large a surface was placed on the cushioning. In addition, some liners, depending on their composition, may react with salts in the soil as well as acid from the chemical leaching to affect the successfulness of the liner. This can be amplified over time.

Environmental Concerns

Effectiveness

Heap leach mining works well for large volumes of low grade ores, as reduced metallurgical treatment (comminution) of the ore is required in order to extract an equivalent amount of minerals when compared to milling. The significantly reduced processing costs are offset by the reduced yield of usually approximately 60-70%. The amount of overall environmental impact caused by heap leaching is often lower than more traditional techniques. It also requires less energy consumption to use this method, which many consider to be an environmental alternative.

Government Regulation

In the United States, the General Mining Law of 1872 gave rights to explore and mine on public domain land; the original law did not require post-mining reclamation (Woody et al. 2011). Mined land reclamation requirements on federal land depended on state requirements until the passage of the Federal Land Policy and Management Act in 1976. Currently, mining on federal land must have a government-approved mining and reclamation plan before mining can start. Reclamation bonds are required. Mining on either federal, state, or private land is subject to the requirements of the Clean Air Act and the Clean Water Act.

One solution proposed to reclamation problems is the privatization of the land to be mined.

Cultural and Social Concerns

With the rise of the environmentalist movement has also come an increased appreciation for social justice, and mining has showed similar trends lately. Societies located near potential mining sites are at increased risk to be subjected to injustices as their environment is affected by the changes made to mined lands—either public or private—that could eventually lead to problems in social structure, identity, and physical health. Many have argued that by cycling mine power through local citizens, this disagreement can be alleviated, since both interest groups would have shared and equal voice and understanding in future goals. However, it is often difficult to match corporate mining interests with local social interests, and money is often a deciding factor in the successes of any disagreements. If communities are able to feel like they have a valid understanding and power in issues concerning their local environment and society, they are more likely to tolerate and encourage the positive benefits that come with mining, as well as more effectively promote alternative methods to heap leach mining using their intimate knowledge of the local geography.

Examples of Rum Jungle Mine

One of the oldest and most famous uranium mines in the world, the Rum Jungle Mine in Northern Territory, Australia was constructed in the 1950s and is today experiencing extreme amounts of environmental degradation and acid rock drainage that are leading to further negative impacts on the surrounding river and ecosystems. unreliable source This mine includes three overburdened heaps, two flooded open cuts, and a backfilled open cut, as well as numerous former tailings and heap leach pads. These leach pads caused considerable soil contamination as chemicals seeped through them. There was attempted rehabilitation in the 1980s, but there are still high evidences of environmental problems today. These waste sites caused the local river to maintain water that is highly unsafe due to its acidity, high concentrations of target minerals, and other toxic chemicals, many of which are said to have originally leached out of. While the highest concentrations have stayed near the buffer zones of the mine and in the East Finniss River, those that did make it into the Finniss River pose a serious and ongoing public threat to those living nearby who used the river daily. Now years later, it is still posing a serious environmental risk to those around the mine.

Ranger Uranium Mine

The Ranger Uranium Mine in Northern Australia showed a significant increase in erodibility of lands when in contact with materials treated from chemical mining. This could manifest itself in landslides and loss of habitat as well as an increase in gravel composition that could cause other potential problems. If these lands can be planted with vegetation that can survive more acidic conditions, however likely that may be, they may be able to avoid eroding materials into other, less mine-contaminated ecosystem. Given the likelihood of this, though, more rehabilitation measures are being sought after.

The Australian government, which deals with negative environmental effects from many historic mines, now requires steps to address environmental and social concerns. Alternative locations must be analyzed in mining proposals, as well as rehabilitation plans, externalities (and possible solutions), groundwater and infrastructure changes, gained or lost opportunities, socio-economic impacts, and risks, as well as measures taken to reduce or eliminate the risks.

Fort Belknap and Montana

Located on the Fort Belknap Indian Reservation, the Zortman-Landusky gold mine in Montana was one of many early heap leach mines that experienced problems with spills and contamination of surface and groundwater. Although the leaks happened in the 1980s, and the mine was eventually shut down in 1996, health problems on the reservation persist, and as not all of the mine was properly cleaned up, could potentially cause further damage to the people of Fort Belknap. Zortman-Landusky eventually filed for bankruptcy when the Bureau of Land Management stepped in to assist the lawsuit that was not heard by the residents of the reservation. Once the bankruptcy was filed, however, all health care and studies ceased, and compensation for the destruction of culturally significant mountain peaks to the local Assiniboine and Gros Ventre people was never achieved. Today, there are still abnormally high reports of health problems including thyroid problems, lead poisoning, chemical burns, and emphysema, especially in children.

In-situ Leaching

In-situ leaching (ISL), also called in-situ recovery (ISR) or solution mining, is a mining process used to recover minerals such as copper and uranium through boreholes drilled into a deposit, *in situ*. In situ leach works by artificially dissolving minerals occurring naturally in a solid state. For recovery of material occurring naturally in solution.

Remains of uranium in-situ leaching in Stráž pod Ralskem, Czech Republic.

The process initially involves the drilling of holes into the ore deposit. Explosive or hydraulic fracturing may be used to create open pathways in the deposit for solution to penetrate. Leaching solution is pumped into the deposit where it makes contact with the ore. The solution bearing the dissolved ore content is then pumped to the surface and processed. This process allows the extraction of metals and salts from an ore body without the need for conventional mining involving drill-and-blast, open-cut or underground mining.

Process

In-situ leach mining involves pumping of a lixiviant into the ore body via a borehole, which circulates through the porous rock dissolving the ore and is extracted via a second borehole.

The lixiviant varies according to the ore deposit: for salt deposits the leachate can be fresh water into which salts can readily dissolve. For copper, acids are generally needed to enhance solubility of the ore minerals within the solution. For uranium ores, the lixiviant may be acid or sodium bicarbonate.

Minerals

Potash and Soluble Salts

In-situ leach is widely used to extract deposits of water-soluble salts such as potash (sylvite and carnallite), rock salt (halite), sodium chloride, and sodium sulfate. It has been used in the US state of Colorado to extract nahcolite (sodium bicarbonate). In-situ leaching is often used for deposits that are too deep, or beds that are too thin, for conventional underground mining.

Uranium

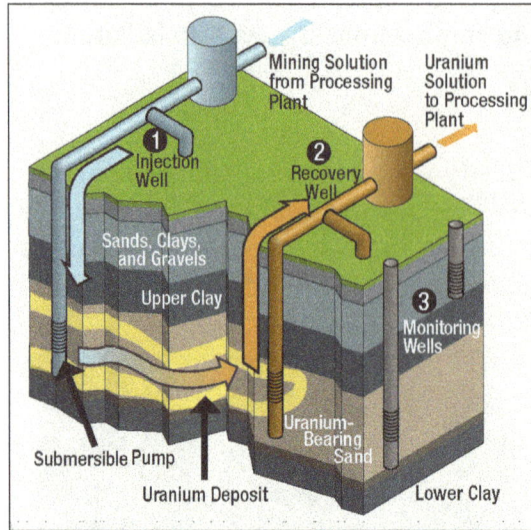

Diagram of in-situ leaching for uranium (US NRC).

In-situ leach for uranium has expanded rapidly since the 1990s, and is now the predominant method for mining uranium, accounting for 45 percent of the uranium mined worldwide in 2012.

Solutions used to dissolve uranium ore are either acid (sulfuric acid or less commonly nitric acid) or carbonate (sodium bicarbonate, ammonium carbonate, or dissolved carbon dioxide). Dissolved oxygen is sometimes added to the water to mobilize the uranium. ISL of uranium ores started in the United States and the Soviet Union in the early 1960s. The first uranium ISL in the US was in the Shirley Basin in the state of Wyoming, which operated from 1961-1970 using sulfuric acid. Since 1970, all commercial-scale ISL mines in the US have used carbonate solutions. ISL mining in Australia uses acid solutions.

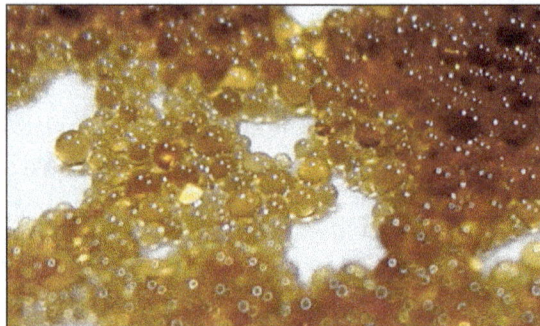

Ion exchange resin beads.

In-situ recovery involves the extraction of uranium-bearing water (grading as low as .05% U_3O_8). The extracted uranium solution is then filtered through resin beads. Through an ion exchange process, the resin beads attract uranium from the solution. Uranium loaded resins are then transported to a processing plant, where U_3O_8 is separated from the resin beads and yellowcake is produced. The resin beads can then be returned to the ion exchange facility where they are reused.

At the end of 2008 there were four in-situ leaching uranium mines operating in the United States, operated by Cameco, Mestena and Uranium Resources, Inc., all using sodium bicarbonate.

ISL produces 90% of the uranium mined in the US. In 2010, Uranium Energy Corporation began in-situ leach operations at their Palangana project in Duval County, Texas. In July 2012 Cameco delayed development of their Kintyre project, due to challenging project economics based on $45.00 U_3O_8. One ISR reclamation project was also in operation as of 2009.

A drum of yellowcake.

Significant ISL mines are operating in Kazakhstan and Australia. The Beverley uranium mine in Australia uses in-situ leaching. ISL mining accounted for 41% of the world's uranium production in 2010.

Examples of in-situ uranium mines include:

- The Beverley Uranium Mine, South Australia, is an operating ISL uranium mine and Australia's first such mine.

- The Honeymoon Uranium Mine, South Australia, opened in 2011 and is Australia's second ISL uranium mine.

- Crow Butte (operating), Smith Ranch-Highland (operating), Christensen Ranch (reclamation), Irigaray (reclamation), Churchrock (proposed), Crownpoint (proposed), Alta Mesa (operating), Hobson (standby), La Palangana (operating), Kingsville Dome (operating), Rosita (standby) and Vasquez (restoration) are ISL uranium operations in the United States.

- In 2010 Uranium Energy Corp. began an ISL mining operation in the Palangana deposit in Duval County, Texas. The ion exchange facility at Palangana trucks uranium-loaded resin beads to the company's Hobson processing plant, where yellowcake is produced. Uranium Energy Corp. has three additional South Texas deposits permitted or in development.

Copper

In-situ leaching of copper was done by the Chinese by 977 AD, and perhaps as early as 177 BC. Copper is usually leached using acid (sulfuric acid or hydrochloric acid), then recovered from solution by solvent extraction electrowinning (SX-EW) or by chemical precipitation.

Ores most amenable to leaching include the copper carbonates malachite and azurite, the oxide tenorite, and the silicate chrysocolla. Other copper minerals, such as the oxide cuprite and the sulfide chalcocite may require addition of oxidizing agents such as ferric sulfate and oxygen to the leachate before the minerals are dissolved. The ores with the highest sulfide contents, such as bornite and chalcopyrite will require more oxidants and will dissolve more slowly. Sometimes oxidation is speeded by the bacteria *Thiobacillus ferrooxidans*, which feeds on sulfide compounds.

Copper ISL is often done by *stope leaching*, in which broken low-grade ore is leached in a current or former conventional underground mine. The leaching may take place in backfilled stopes or caved areas. In 1994, stope leaching of copper was reported at 16 mines in the US.

Recovery well at former San Manuel operation.

At the San Manuel mine in the US state of Arizona, ISL was initially used by collecting the resultant solution underground but in 1995 this was converted to a well-to-well recovery method which was the first large scale implementation of that method. This well-to-well method has been proposed for other copper deposits in Arizona.

Gold

In-situ leaching has not been used on a commercial scale for gold mining. A three-year pilot program was undertaken in the 1970s to in-situ leach gold ore at the Ajax mine in the Cripple Creek district in the US, using a chloride and iodide solution. After obtaining poor results, perhaps because of the complex telluride ore, the test was halted.

Environmental Concerns

According to the World Nuclear Organization:

> "In the USA legislation requires that the water quality in the affected aquifer be restored so as to enable its pre-mining use. Usually this is potable water or stock water (usually less than 500 ppm total dissolved solids), and while not all chemical characteristics can be returned to those pre-mining, the water must be usable for the same purposes as before. Often it needs to be treated by reverse osmosis, giving rise to a problem in disposing of the concentrated brine stream from this".

> "The usual radiation safeguards are applied at an ISL Uranium mining operation, despite the fact that most of the orebody's radioactivity remains well underground and there is hence minimal increase in radon release and no ore dust. Employees are monitored for alpha

radiation contamination and personal dosimeters are worn to measure exposure to gamma radiation. Routine monitoring of air, dust and surface contamination are undertaken".

The advantages of this technology are:

- Reduced hazards for the employees from accidents, dust, and radiation.

- Low cost, no need for large uranium mill tailings deposits.

After termination of an in-situ leaching operation, the waste slurries produced must be safely disposed, and the aquifer, contaminated from the leaching activities, must be restored. Groundwater restoration is a very tedious process that is not yet fully understood.

The best results have been obtained with the following treatment scheme, consisting of a series of different steps:

- Phase 1: Pumping of contaminated water: the injection of the leaching solution is stopped and the contaminated liquid is pumped from the leaching zone. Subsequently, clean groundwater flows in from outside of the leaching zone.

- Phase 2: As 1, but with treatment of the pumped liquid (by reverse osmosis) and re-injection into the former leaching zone. This scheme results in circulation of the liquid.

- Phase 3: As 2, with the addition of a reducing chemical (for example hydrogen sulfide (H_2S) or sodium sulfide (Na_2S). This causes the chemical precipitation and thus immobilization of major contaminants.

- Phase 4: Circulation of the liquid by pumping and re- injection, to obtain uniform conditions in the whole former leaching zone.

But, even with this treatment scheme, various problems remain unresolved:

- Contaminants that are mobile under chemically reducing conditions, such as radium, cannot be controlled.

- If chemically reducing conditions are later disturbed for any reasons, the precipitated contaminants are re-mobilized.

- The restoration process takes very long periods of time, not all parameters can be lowered appropriately.

Most restoration experiments reported refer to the alkaline leaching scheme, since this scheme is the only one used in Western world commercial in-situ operations. Therefore, nearly no experience exists with groundwater restoration after acid in- situ leaching, the scheme that was applied in most instances in Eastern Europe. The only Western in-situ leaching site restored after sulfuric acid leaching so far, is the small pilot scale facility Nine Mile Lake near Casper, Wyoming (USA). The results can therefore not simply be transferred to production scale facilities. The restoration scheme applied included the first two steps mentioned above. It turned out that a water volume of more than 20 times the pore volume of the leaching zone had to be pumped, and still several parameters did not reach background levels. Moreover, the restoration required about the same time as used for the leaching period.

In USA, the Pawnee, Lamprecht, and Zamzow ISL Sites in Texas were restored using steps 1 and 2 of the above listed treatment scheme. Relaxed groundwater restoration standards have been granted at these and other sites, since the restoration criteria could not be met.

A study published by the U.S. Geological Survey in 2009 found that "To date, no remediation of an ISR operation in the United States has successfully returned the aquifer to baseline conditions".

Baseline conditions include commercial quantities of radioactive U_3O_8. Efficient in-situ recovery reduces U_3O_8 values of the aquifer. Speaking at an EPA Region 8 workshop, on September 29, 2010, Ardyth Simmons on the subject "Establishing Baseline and Comparison to Restoration Values at Uranium In-Situ Recovery Sites" stated "These results indicated that it may be unrealistic for ISR operations to restore aquifers to the mean, because in some cases, this means that there would have to be less uranium present than there was pre-mining. Pursuing more conservative concentrations results in a considerable amount of water usage, and many of these aquifers were not suitable for drinking water before mining initiated".

The EPA is considering the need to update the environmental protection standards for uranium mining because current regulations, promulgated in response to the Uranium Mill Tailings Radiation Control Act of 1978, do not address the relatively recent process of in-situ leaching (ISL) of uranium from underground ore bodies. In a February, 2012 letter the EPA states, "Because the ISL process affects groundwater quality, the EPA's Office of Radiation and Indoor Air requested advice from the Science Advisory Board (SAB) on issues related to design and implementation of groundwater monitoring at ISL mining sites".

The SAB makes recommendations concerning monitoring to characterize baseline groundwater quality prior to the start of mining operations, monitoring to detect any leachate excursions during mining, and monitoring to determine when groundwater quality has stabilized after mining operations have been completed. The SAB also reviews the advantages and disadvantages of alternative statistical techniques to determine whether post-operational groundwater quality has returned to near pre-mining conditions and whether mine operation can be predicted not to adversely impact groundwater quality after site closure acceptance.

PYROMETALLURGY

Pyrometallurgy deals with chemical reactions at high temperatures (ranging from 100°C up to 3000°C). These reactions involve numerous different solids, liquids, and gases, and are carried out using many diverse types of furnaces.

Principles of Production of Metals by Pyrometallurgy

In pyrometallurgy, metals are extracted by converting sulfides into oxides and then reducing oxides into metals, using carbon or carbon monoxide as reducing agents. The reactions for lead, zinc, and iron are given below. The reader is cautioned that the following chemical reactions are overly simplified versions of the actual processes. In these reactions, $CaCO_3$ is a flux.

Lead:

$$2\,PbS + 3\,O_2 \Longrightarrow 2\,PbO + 2\,SO_2$$

$$CaCO_3 \Longrightarrow CaO + CO_2$$

$$C + CO_2 \Longrightarrow 2\,CO$$

$$PbO + CO \Longrightarrow Pb + CO_2$$

Zinc:

$$2\,ZnS + 3\,O_2 \Longrightarrow 2\,ZnO + 2\,SO_2$$

$$CaCO_3 \Longrightarrow CaO + CO_2$$

$$C + CO_2 \Longrightarrow 2\,CO$$

$$ZnO + CO \Longrightarrow Zn + CO_2$$

Iron:

$$3\,Fe_2O_3 + CO \Longrightarrow 2\,Fe_3O_4 + CO_2$$

$$Fe_3O_4 + CO \Longrightarrow 3\,FeO + CO_2$$

$$FeO + CO \Longrightarrow Fe + CO2$$

Heat of Reaction (Enthalpy)

The heat of reaction (ΔH), or enthalpy, determines the energy cost of the process. If the reaction is exothermic (ΔH is negative), then heat is given off by the reaction, and the process will be partially self-heating. If the reaction is endothermic (ΔH is positive), then the reaction absorbs heat, which will have to be supplied to the process.

Equilibrium

Most of the reactions used in pyrometallurgy are reversible, and so they will reach an equilibrium where the desired products are converting back into the reactants as quickly as the reactants are forming the products:

$$A + B <===> C + D$$

We would prefer the reaction:

$$A + B \Longrightarrow C + D$$

and so the process will need to remove C and D as quickly as they are produced, so that the products cannot react to re-form A and B. A gaseous product can be removed by venting it off, and other types of product can be removed by dissolving them in slags of an appropriate composition.

Molten metals tend to dissolve impurities from the ore. For example, many ores that contain native copper have arsenic compounds associated with them. When the ore is melted, the molten copper dissolves the arsenic. As a result, the metal from the furnace can be less pure than the individual metal grains that were originally in the ground.

Gibbs Free Energy

The Gibbs free energy (ΔG) of a reaction is a measure of the thermodynamic driving force that makes a reaction occur. A negative value for ΔG indicates that a reaction can proceed spontaneously without external inputs, while a positive value indicates that it will not. The equation for Gibbs free energy is:

$$\Delta G = \Delta H - T\Delta S$$

where ΔH is the enthalpy change in the reaction, T is absolute temperature, and ΔS is the entropy change in the reaction.

The enthalpy change (ΔH) is a measure of the actual energy that is liberated when the reaction occurs (the "heat of reaction"). If it is negative, then the reaction gives off energy, while if it is positive the reaction requires energy.

The entropy change (ΔS) is a measure of the change in the possibilities for disorder in the products compared to the reactants. For example, if a solid (an ordered state) reacts with a liquid (a somewhat less ordered state) to form a gas (a highly disordered state), there is normally a large positive change in the entropy for the reaction.

Construction of an Ellingham Diagram

An Ellingham diagram is a plot of ΔG versus temperature. Since ΔH and ΔS are essentially constant with temperature unless a phase change occurs, the free energy versus temperature plot can be drawn as a series of straight lines, where ΔS is the slope and ΔH is the y-intercept. The slope of the line changes when any of the materials involved melt or vaporize.

Free energy of formation is negative for most metal oxides, and so the diagram is drawn with $\Delta G=0$ at the top of the diagram, and the values of ΔG shown are all negative numbers.

The Ellingham diagram is for metals reacting to form oxides (similar diagrams can also be drawn for metals reacting with sulfur, chlorine, etc., but the oxide form of the diagram is most common). The oxygen partial pressure is taken as 1 atmosphere, and all of the reactions are normalized to consume one mole of O_2.

The majority of the lines slope upwards, because both the metal and the oxide are present as condensed phases (solid or liquid). The reactions are therefore reacting a gas with a condensed phase to make another condensed phase, which reduces the entropy. A notable exception to this is the oxidation of solid carbon. The line for the reaction:

$$C + O_2 ==> CO_2$$

is a solid reacting with a mole of gas to produce a mole of gas, and so there is little change in entropy and the line is nearly horizontal. For the reaction:

$$2C + O_2 \Longrightarrow 2CO$$

we have a solid reacting with a gas to produce two moles of gas, and so there is a substantial increase in entropy and the line slopes rather sharply downward. Similar behavior can be seen in parts of the lines for lead and lithium, both of which have oxides that boil at slightly lower temperatures than the metal does.

There are three main uses of the Ellingham diagram:

- Determine the relative ease of reducing a given metallic oxide to metal;

- Determine the partial pressure of oxygen that is in equilibrium with a metal oxide at a given temperature; and

- Determine the ratio of carbon monoxide to carbon dioxide that will be able to reduce the oxide to metal at a given temperature.

Ease of Reduction

The position of the line for a given reaction on the Ellingham diagram shows the stability of the oxide as a function of temperature. Reactions closer to the top of the diagram are the most "noble" metals (for example, gold and platinum), and their oxides are unstable and easily reduced. As we move down toward the bottom of the diagram, the metals become progressively more reactive and their oxides become harder to reduce.

A given metal can reduce the oxides of all other metals whose lines lie above theirs on the diagram. For example, the $2Mg + O_2 \Longrightarrow 2MgO$ line lies below the $Ti + O_2 \Longrightarrow TiO_2$ line, and so metallic magnesium can reduce titanium oxide to metallic titanium.

Since the $2C + O_2 \Longrightarrow 2CO$ line is downward-sloping, it cuts across the lines for many of the other metals. This makes carbon unusually useful as a reducing agent, because as soon as the carbon oxidation line goes below a metal oxidation line, the carbon can then reduce the metal oxide to metal. So, for example, solid carbon can reduce chromium oxide once the temperature exceeds approximately 1225°C, and can even reduce highly-stable compounds like silicon dioxide and titanium dioxide at temperatures above about 1620°C and 1650°C, respectively. For less stable oxides, carbon monoxide is often an adequate reducing agent.

Equilibrium Partial Pressure of Oxygen

The scale on the right side of the diagram labelled "P_{O_2}" is used to determine what partial pressure of oxygen will be in equilibrium with the metal and metal oxide at a given temperature. The significance of this is that, if the oxygen partial pressure is higher than the equilibrium value, the metal will be oxidized, and if it is lower than the equilibrium value then the oxide will be reduced.

To use this scale, you will need a straightedge. First, find the temperature you are interested in, and find the point where the oxidation line of interest crosses that temperature. Then, line up the straightedge with both that point, and with the point labelled "o" that is marked with short radiating lines (upper left corner of the diagram). Now, with the straightedge running through these two points, read off the oxygen partial pressure (in atmospheres) where the straightedge crosses the "Po_2" scale, and this is the equilibrium partial pressure.

It is possible to reach the equilibrium oxygen partial pressure by use of a hard vacuum, by purging with an inert gas to displace the oxygen, or by using a scavenger chemical to consume the oxygen.

Ratio of CO/CO_2 Needed for Reduction

When using carbon as a reducing agent, there will be a minimum ratio of CO to CO_2 that will be able to reduce a given oxide. The harder the oxide is to reduce, the greater the proportion of CO needed in the gases.

To determine the CO/CO_2 ratio to reduce a metal oxide at a particular temperature, use the same procedure as for determining the equilibrium pressure of oxygen, except line up the straightedge with the point marked "C" (center of the left side of the diagram), and read the ratio off of the scale marked "CO/CO_2".

Pyrometallurgical Equipment

Types of Furnaces

1. Shaft Furnace: These are vertical furnaces with the charge added at the top and removed at the bottom, while gas is blown into the bottom and exits the top, as shown in figure. The solid charge must consist of particles coarse enough that they will not be blown out of the furnace by the gas. An iron ore blast furnace is a typical example of a shaft furnace.

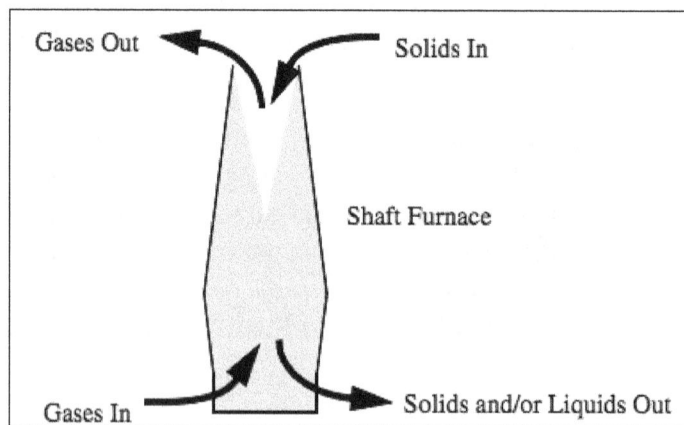

Schematic of a shaft furnace, showing the material flows.

2. Muffle Furnace: This type of furnace is used when the material being heated should not be contaminated by the heating fuel. This is accomplished by enclosing the material in a chamber, with the fuel burned outside of the chamber.

3. Hearth Furnace: Hearth furnaces allow the burning fuel to come in contact with the material being heated. This allows very high temperatures to be reached. This type of furnace includes reverberatory furnaces and rotary kilns.

4. Electric Furnace: Electric furnaces heat the charge by running a massive electrical current through it (the larger furnaces use approximately 20,000 amps at 50 to 500 volts). These work by immersing electrodes in the charge.

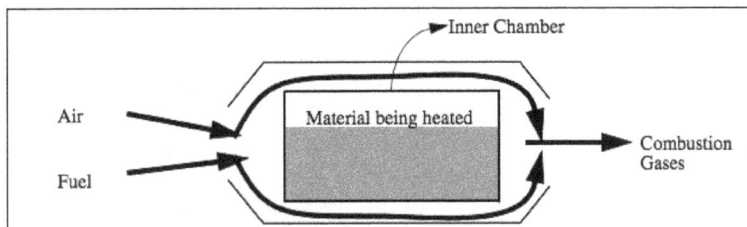

Schematic of a muffle furnace.

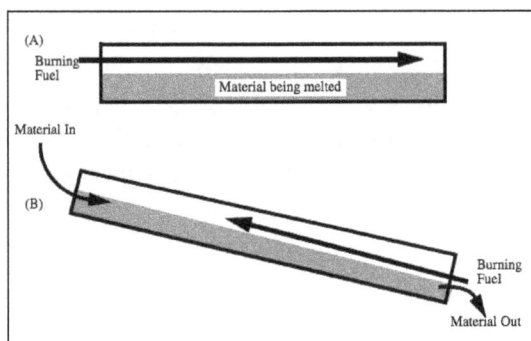

Hearth furnaces. (A) Reverberatory furnace; (B) Rotary Kiln.

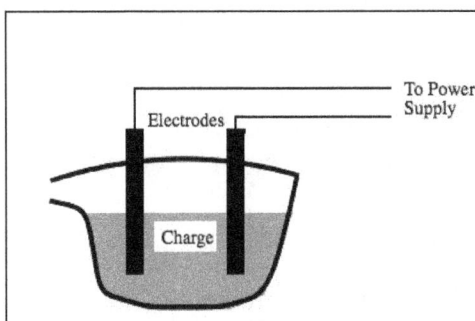

Schematic of an electric furnace. The electrodes can be connected to DC, AC, or 3-phase AC, depending on the application.

Drying and Calcination

Drying

Drying is the removal of water from the ore, using a moderate amount of heat (temperatures on the order of 100°C). Only the mechanically bound water is removed (water filling pores and cracks, or that is adhering to the particle surfaces). Chemically-bound water, such as water of hydration in the ore crystal structure, is not removed by drying processes.

Calcination

Originally, calcination referred to the heating of limestone above 900°C to drive off the CO_2 and produce lime:

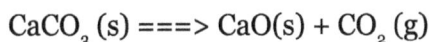

$$CaCO_3 \text{ (s)} ===> CaO\text{(s)} + CO_2 \text{ (g)}$$

In current practice, calcination refers to any process where the material is heated to drive off volatile organics, CO_2, chemically bound water, or similar compounds. For example:

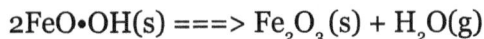

$$2Al(OH)_3 \text{ (s)} ===> Al_2O_3 \text{ (s)} + 3H_2O\text{(g)}$$

$$2FeO \cdot OH\text{(s)} ===> Fe_2O_3 \text{ (s)} + H_2O\text{(g)}$$

Roasting

Roasting involves not only heating, but also reaction with a gas. It is typically used to convert sulfides to oxides by reaction with air (air is usually used as an oxidizing agent, because it is free). For example:

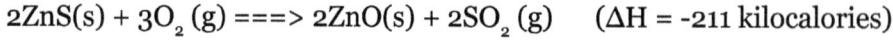

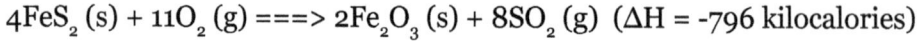

$$2ZnS(s) + 3O_2 (g) ===> 2ZnO(s) + 2SO_2 (g) \quad (\Delta H = -211 \text{ kilocalories})$$

$$4FeS_2 (s) + 11O_2 (g) ===> 2Fe_2O_3 (s) + 8SO_2 (g) \quad (\Delta H = -796 \text{ kilocalories})$$

Roasting can also use other gases, such as chlorine (to produce volatile chlorides):

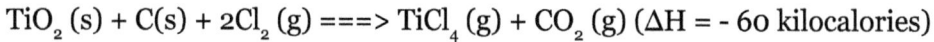

$$TiO_2 (s) + C(s) + 2Cl_2 (g) ===> TiCl_4 (g) + CO_2 (g) (\Delta H = -60 \text{ kilocalories})$$

If the reaction is exothermic enough (strongly negative ΔH), then autogenous roasting may be possible. This is where the heat for the roast is provided by the roasting reaction, and fuel is only needed to get the reaction started. For this to work, the furnace must be designed to capture and use the heat produced to bring fresh ore up to the roasting temperature. Not all ores will "burn" in this way, and so in many cases supplemental fuel will be needed to maintain the roasting temperature.

Basic Steps in Roasting

- Particles are heated.
- Reactive gas (air, oxygen, chlorine, etc.) contacts the particles.
- Particles react with the gas.
- Gaseous reaction product are carried away.

Since the particles do not melt, the reaction starts on the particle surface and gradually works in to the particle core, as shown in the Shrinking-Core reaction model.

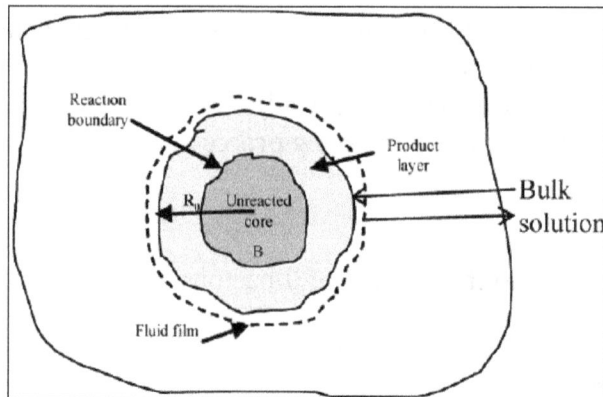

The Shrinking-Core model. As the shell of oxidized ore becomes thicker, it becomes more difficult for fresh gas to reach the unreacted ore, and so the roasting rate slows down. It is often difficult to react the last bit of material in the center of the particles.

Basic Roasting Terms

- Dead Roast: the ore is completely reacted, and leaves the process cold.

- Sweet Roast: the ore is completely reacted, but leaves the furnace still hot.

- Sour Roast: the roasting reaction is not run to completion.

Types of Roasting Furnaces

Hearth Roaster: This type of furnace is suitable for coarse particles, and is common in older facilities. The capacity is on the order of 100-200 tons/day.

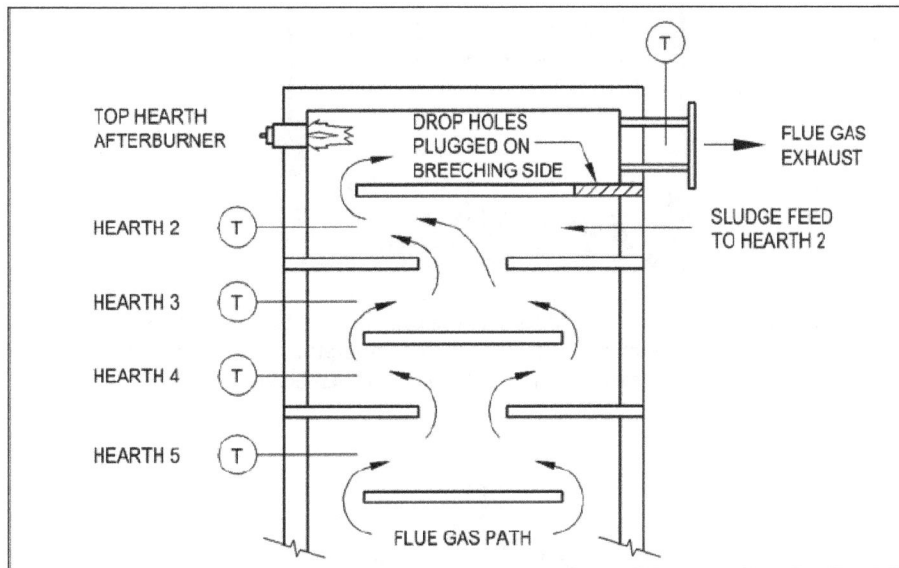

Hearth Roaster. Typically, these have 6 to 12 hearths, and most of the roasting occurs as the particles drop from one hearth to the next.

Flash or Suspension Roaster: These furnaces process very fine particles, and take advantage of the rapid reaction rate to provide autogenous heating. The capacity is approximately 3 to 4 times that of a similar-sized hearth roaster.

Fluidized Bed Roaster: Fluidized bed reactors are used not only for roasting, but also for drying and calcination. Gas is bubbled up through a bed of particles, with the particles large enough that they are not swept out of the furnace, but small enough that the gas can expand the bed so that it will behave as a fluid. These reactors provide excellent gas/solid contact, without requiring extremely fine particles.

Smelting

In smelting, the ore is brought to a high enough temperature that the material melts, and the final product is a molten metal and a slag. Smelting is done so that the impurities are either carried off in the slag, or are burned off as a gas. In some types of furnace (blast furnaces, flash smelters) the roasting and smelting operations are combined.

Blast Furnaces: The function of a blast furnace is to produce pig iron, not steel. A blast furnace is a shaft furnace in which a blast of preheated air is blown in through tuyeres at the bottom. A schematic of a blast furnace used for making iron is shown in figure shows a description of the reactions that take place in the furnace.

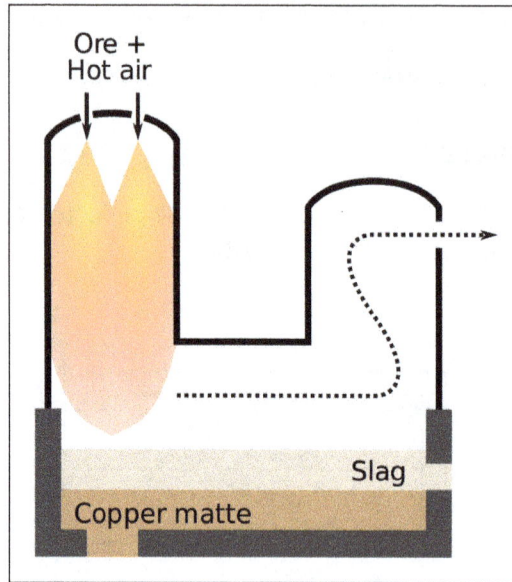

Flash Smelting. The heat produced by the roasting reaction is used to dry the incoming ore and bring it up to roasting temperature. If the reaction is sufficiently exothermic, the ore can even be melted.

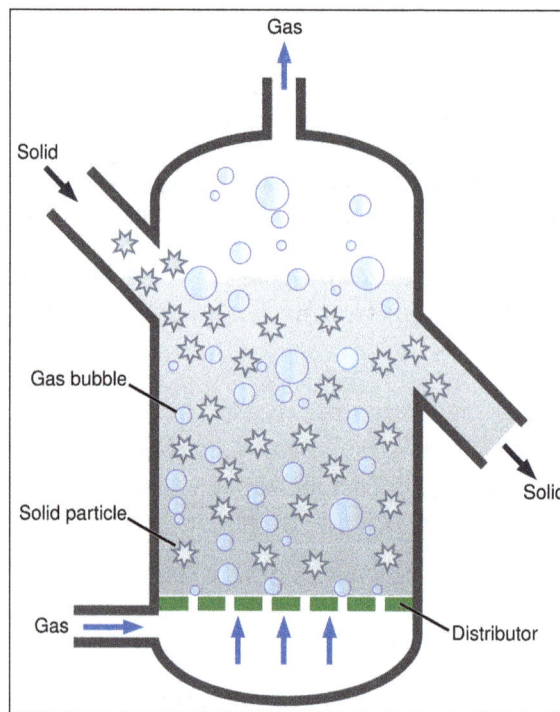

Fluidized Bed Reactor.

Components of the charge to a blast furnace:

- Iron Ore.

- Coke.

- Fluxes.

Iron Ore

Iron ore can be hematite (Fe_2O_3), magnetite (Fe_3O_4) or taconite (a colloquial term used for Minnesota ores which can refer to either hematite or magnetite). Iron ore is beneficiated to 65 - 72% Fe, and pelletized to form pellets 3/8 inch to 5/8 inch in diameter. In some operations, concentrated fines may also be sintered, that is, fused into porous lumps that are broken into one or two inch pieces. Pellets are durable and ship well, and sinter does not. Pelletizing plants are therefore often built near the mine and pellets are transported to the blast furnace by rail or ship, while sinter is usually produced at the steel mill.

Coke

Coke serves three functions:

- Supply chemical reactants for reducing iron ore to metallic iron.

- Act as a source of carbon in the pig iron and eventually in the steel.

- Provide a source of heat in the blast furnace (fuel).

Fluxes

When metal is smelted, the metal is separated from its impurities by melting, with the impurities forming a molten slag on top of the metal. Many of the impurities associated with iron ore are difficult to melt, and so they will not form a proper slag easily, which retards the smelting process. To make these impurities easier to melt, fluxes are added.

Limestone ($CaCO_3$) or dolomite ($(Ca,Mg)CO_3$) are two typical fluxes used in blast furnaces. When a large amount of sulfur needs to be removed from the furnace charge, limestone is the preferred flux. Limestone is also a better flux to use if slag from the blast furnace is to be used as a raw material for cement manufacture. An important criterion for flux selection is availability and cost, and dolomite is often more readily available and less expensive than limestone.

Components of a Blast Furnace

The components of a blast furnace are shown in figure, and brief descriptions of the functions of each are as follows:

Stack: In the blast-furnace stack, there is a countercurrent flow of gas and solids. In the stack, hematite or magnetite is reduced to sponge iron by the action of carbon monoxide.

Bosh: This is the part of the furnace where the contents melt. As shown in figure, several chemical reactions take place in the bosh. This is where the slag is formed.

Hearth: In the hearth, molten pig iron and molten slag segregate from each other and are tapped off separately. Slag is lower density than iron, and so it floats on the molten metal.

Height: 318 feet
Diameter: 45 feet

Hopper

Flue

Bell

Wear Plates

Top

Stack

Top view of furnace

Bustle pipe

Tuyeres

Furnace

Inwall

Bustle Pipe

Bosh

Tuyere

Cinder Notch

Mantle

Hearth

Tap Hole

Paved Floor

Bottom

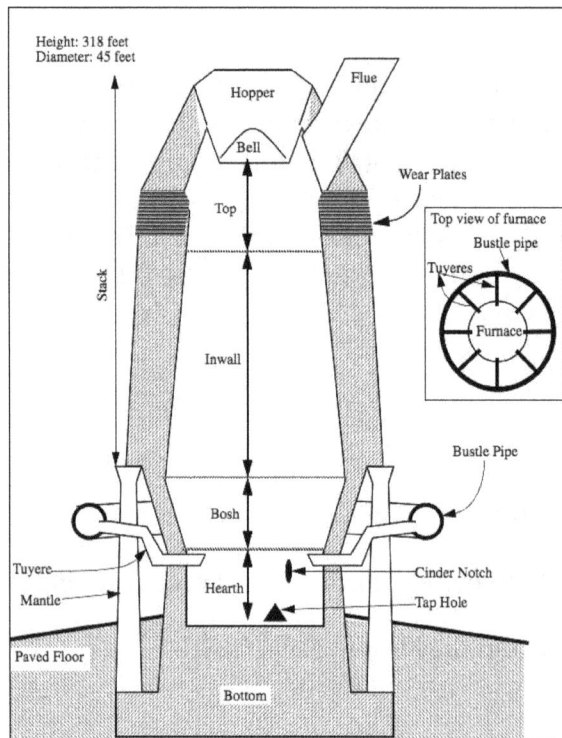

The iron blast furnace. Heated air is pumped into the bustle pipe, and injected into the furnace through the tuyeres. The bell acts as an air-lock, so that the ore charge can be added to the furnace without letting all the gas escape.

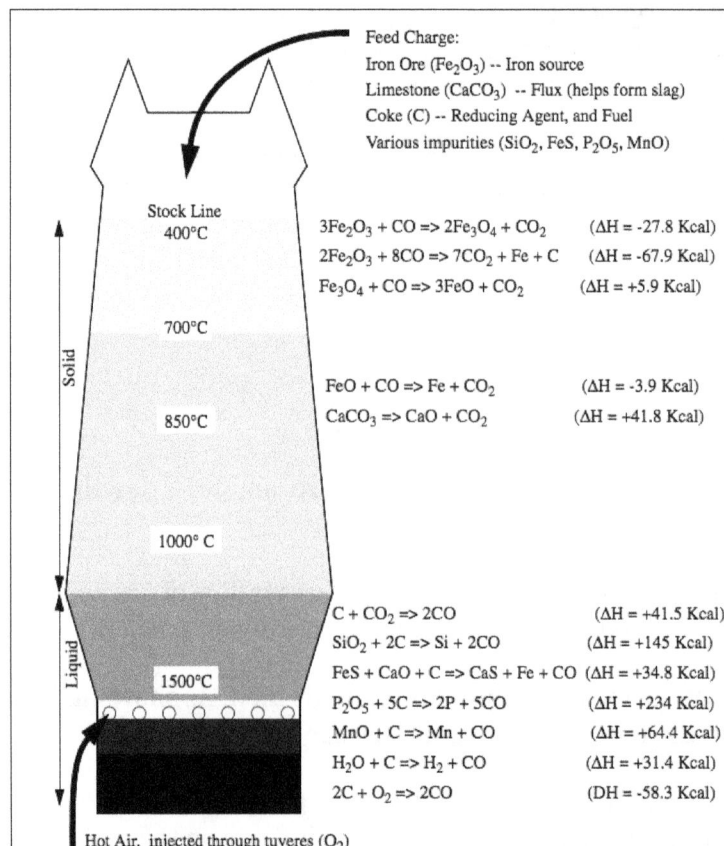

Feed Charge:
Iron Ore (Fe_2O_3) -- Iron source
Limestone ($CaCO_3$) -- Flux (helps form slag)
Coke (C) -- Reducing Agent, and Fuel
Various impurities (SiO_2, FeS, P_2O_5, MnO)

Stock Line
400°C

$3Fe_2O_3 + CO \Rightarrow 2Fe_3O_4 + CO_2$ ($\Delta H = -27.8$ Kcal)
$2Fe_2O_3 + 8CO \Rightarrow 7CO_2 + Fe + C$ ($\Delta H = -67.9$ Kcal)
$Fe_3O_4 + CO \Rightarrow 3FeO + CO_2$ ($\Delta H = +5.9$ Kcal)

700°C

Solid

$FeO + CO \Rightarrow Fe + CO_2$ ($\Delta H = -3.9$ Kcal)
$CaCO_3 \Rightarrow CaO + CO_2$ ($\Delta H = +41.8$ Kcal)

850°C

1000° C

$C + CO_2 \Rightarrow 2CO$ ($\Delta H = +41.5$ Kcal)
$SiO_2 + 2C \Rightarrow Si + 2CO$ ($\Delta H = +145$ Kcal)
$FeS + CaO + C \Rightarrow CaS + Fe + CO$ ($\Delta H = +34.8$ Kcal)
$P_2O_5 + 5C \Rightarrow 2P + 5CO$ ($\Delta H = +234$ Kcal)
$MnO + C \Rightarrow Mn + CO$ ($\Delta H = +64.4$ Kcal)
$H_2O + C \Rightarrow H_2 + CO$ ($\Delta H = +31.4$ Kcal)
$2C + O_2 \Rightarrow 2CO$ (DH = -58.3 Kcal)

Liquid

1500°C

Hot Air, injected through tuyeres (O_2)

Temperatures and reactions in the iron blast.

Blast Furnace Operation

The furnace is charged with iron ore lumps, pellets, and sinter; coke; and possibly extra flux. These are carried to the top of the furnace with skips or conveyors, and are tipped, or charged, into the furnace. Meanwhile, air preheated to 900°C is injected through the tuyeres, which are nozzles at the bottom of the furnace. The coke is partially burned by the injected air both to produce heat, and to generate carbon monoxide. Since coke is relatively expensive, some furnaces inject coal or oil along with the air as supplemental fuels to reduce coke usage. The carbon monoxide travels upward through the column, and removes oxygen from the iron ores on their way down, leaving metallic iron. By the time the charge reaches the base of the furnace, the heat generated there melts the iron. The resulting molten "hot metal" is tapped at regular intervals by opening the "tap hole" in the bottom of the furnace so that it can flow out. The fluxes combine with impurities in the coke and ore to form the slag, which floats on the iron and is removed through the "cinder notch".

The hot metal from the furnace is collected in specially-constructed railway containers, called "torpedo cars". The torpedo cars carry the molten iron to the steelmaking furnace.

Blast furnaces are operated continuously without shutdown for ten years or more. If the furnace were allowed to cool, thermal stresses can cause damage to the refractory bricks. Eventually, the refractory bricks in the furnace will wear away, and at that point the furnace is emptied and shut down so that it can be relined with new bricks. The period between shutdowns is referred to as a "campaign".

Iron taken directly from the blast furnace contains about 4 - 4.5% carbon, as well as a number of other elements. This is referred to as "pig iron", and if it is allowed to solidify it is brittle, difficult to work with, and has poor structural properties. The pig iron can be converted to steel by refining in the steelmaking process, which reduces the carbon content and removes other impurities, to make a stronger, tougher, and more generally useful product.

Steel Production

In the United States, we produce about 100 million tons of steel per year. Steel is a very valuable commodity.

Blast furnaces do not produce steel, they produce pig iron. The differences between pig iron and steel composition.

Table: Composition differences between pig iron and steel.

	Pig Iron	Steel
Carbon	3-4%	0.04-1.7%
Manganese	0.15-2.5%	0.3-0.9%
Silicon	0.5-4%	Trace
Sulfur	Up to 0.2%	0.02-0.04%
Phosphorus	0.025-2.5%	<0.04%

Steel is made in two different types of facilities:

- Integrated steel mills (blast furnace operations using iron ore as feed).

- Minimills (electric arc furnace operations using steel scrap as feed).

Integrated Steel Mills

In the integrated steel mills, molten pig iron is produced in a blast furnace and transported to a Basic Oxygen Furnace (BOF) or Basic Oxygen Process (BOP) vessel, or a related unit. In the oxygen steelmaking process, high purity oxygen is blown under pressure onto or over a bath containing hot metal.

The basic steelmaking reactions with this type of furnace are:

- Carbon reacts with oxygen to form CO and CO_2, which escape as gases.

- Silicon, Manganese, and Phosphorus react with oxygen to form their oxides, which are removed in the slag.

- Sulfur reacts with calcium (from the flux added to the furnace) to form calcium sulfide, and is removed in the slag.

The progress of refining in a top-blown basic-oxygen furnace is shown in figure. This is a considerable improvement over the old open-hearth furnace process, which took several hours to accomplish what the basic-oxygen furnace can do in a matter of minutes.

The molten steel is sampled at intervals to determine its carbon content and content of other elements. When the desired composition is reached, the vessel is rotated to pour off the molten steel.

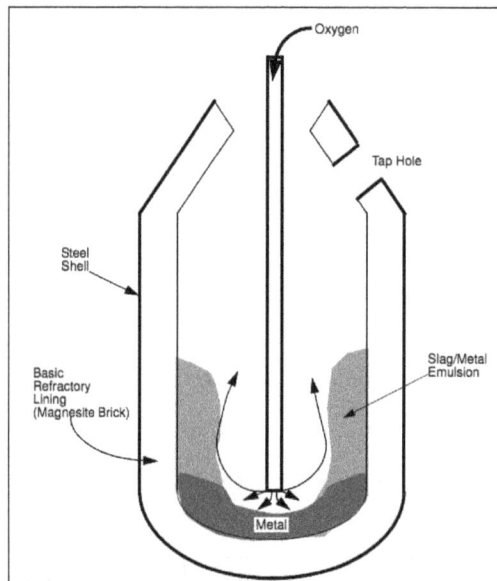

Basic Oxygen Furnace cross section. The oxygen is blown into the furnace to burn off excess carbon, and to convert manganese, silicon, sulfur, and phosphorus into oxides that will segregate into the slag.

The properties of steel depend not only on its chemical composition, but also on the heat treatment. At high temperatures, iron and carbon in steel combine to form iron carbide (Fe_3C), which is commonly known as cementite.

$$3Fe + C ===> Fe_3C$$

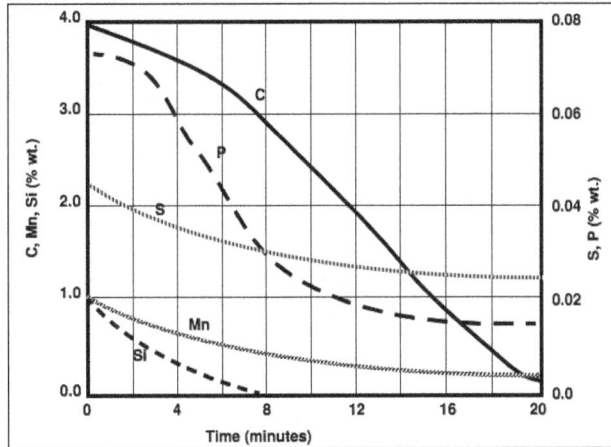

Progress of refining in a top-blown vessel.

The forward reaction is endothermic, so that the formation of cementite is favored at high temperatures. When the steel is cooled slowly, the equilibrium shifts back and the cementite begins to decompose back into iron and small particles of graphite. Steel that contains graphite tends to be gray, while steel that contains cementite is lighter colored, harder, and more brittle.

Heat treatment of the steel controls the ratios of graphite to cementite, and also causes other crystallization effects that allow the mechanical properties of steel to be altered over a very wide range.

Secondary Steelmaking

In the integrated steel mills, an increasing proportion of the steels undergo what is called "secondary steelmaking" before they are cast. The objective of secondary steelmaking is to finetune the chemical composition of the steel, improve its homogeneity, and remove residual impurities. This is done by first tapping the molten steel from the furnace into a ladle, which is fitted with a lid to conserve heat. The ladle of steel can then be treated by any of several different processes, including argon stirring, addition of alloying elements, vacuum degassing, and powder injection. If necessary, an electric arc can be used to keep the molten steel at the proper temperature for casting.

Integrated Steelmaking

The first step in integrated steelmaking is mining the ore. The second step is preparing the raw materials. Ores must be beneficiated to increase the iron content to 65-72% Fe. If the ore is produced far away from the steel mills, then the concentrated ore is mixed with a binder and formed into pellets with diameters between 3/8 inch and 5/8 inch. These pellets are easy to handle and ship well for long distances. If the iron-bearing fines are produced very close to the steel mill, then they can be sintered, that is, fused into porous lumps that are broken into pieces 1 - 2 inches in diameter. Sinter does not ship well, and so a sintering plant is usually a part of the integrated steel mill, while pelletization plants are usually located at the mine.

Preparation of raw materials also includes limestone crushing to a convenient size, and production of coke from coal. Coke is made by heating coal in the absence of oxygen, which drives off the volatiles and leaves a hard, porous, high-carbon coke that is crushed to an appropriate size and charged to the furnace.

The third step in an integrated steel mill is ironmaking. The prepared ore, coke, and limestone are charged into the top of a blast furnace in the correct ratio, while a blast of hot air is injected at the bottom of the furnace. The heated air burns the coke to carbon monoxide, which rises through the downward-moving ore to reduce the iron oxides to metallic iron. When the iron reaches the bottom of the furnace, the temperature is sufficient to melt the iron, producing liquid pig iron.

In the fourth step, the liquid pig iron is transferred to a a steelmaking furnace, usually a basicoxygen furnace (BOF) or an electric-arc furnace (EAF). In this step, the metal is refined by removing unwanted elements into a slag, and adjusting the concentrations of desirable elements. In particular, the carbon content is brought down to between 1.7% and 0.3%.

In the final step, the molten "raw" steel is poured into a ladle, where it may undergo secondary steelmaking processes before being poured into molds where it solidifies into ingots or slabs of steel that can weigh 60 - 100 tons.

Electric Arc Furnace Minimills

Small steel mills, which use electric arc furnaces to melt scrap steel, are referred to as "minimills". A schematic of a minimill flowsheet is included in figure. Electric arc furnaces consist of a tiltable vessel with a removable lid. The lid contains electrodes which are lowered into the furnace and contact the scrap steel charge. An electric current is passed through the electrodes to form an arc. The heat generated by the arc rapidly melts the scrap.

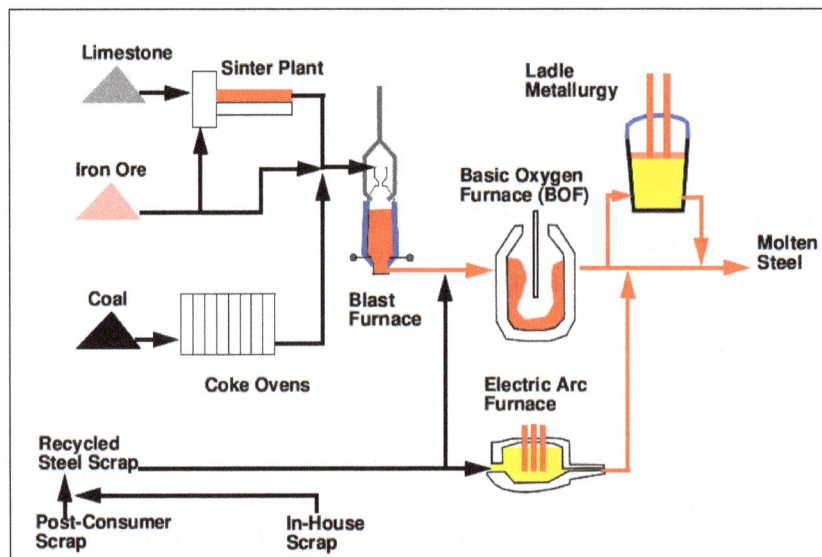

Steelmaking by both the integrated steel mill route, and by the electric-arc furnace minimill route.

During the melting process, other metals are added to the steel to give it the desired chemical composition. As with the basic oxygen furnace, oxygen can be blown into the electric arc furnace to purify the steel. Fluxing agents such as limestone and fluorspar are added to combine with impurities and form a slag.

Once steel with satisfactory chemical composition is formed, the furnace is tilted to pour off the slag which floats on top of the molten steel. After the slag is removed, the furnace is tilted the other direction to pour the molten steel into a ladle. The steel is then transported either to secondary steelmaking, or cast into slabs or ingots.

A typical modern electric arc furnace melts 150 tons of steel in a charge, and a melting cycle takes approximately 90 minutes.

Primary Copper Production

The traditional method for producing metallic copper is to use high-temperature pyrometallurgical techniques to smelt copper sulfides. In the last century, low-temperature hydrometallurgical techniques have been developed which have been very complimentary to the conventional smelting methods, as they allow the treatment of the low-grade copper oxide ores that often occur along with copper sulfide ores.

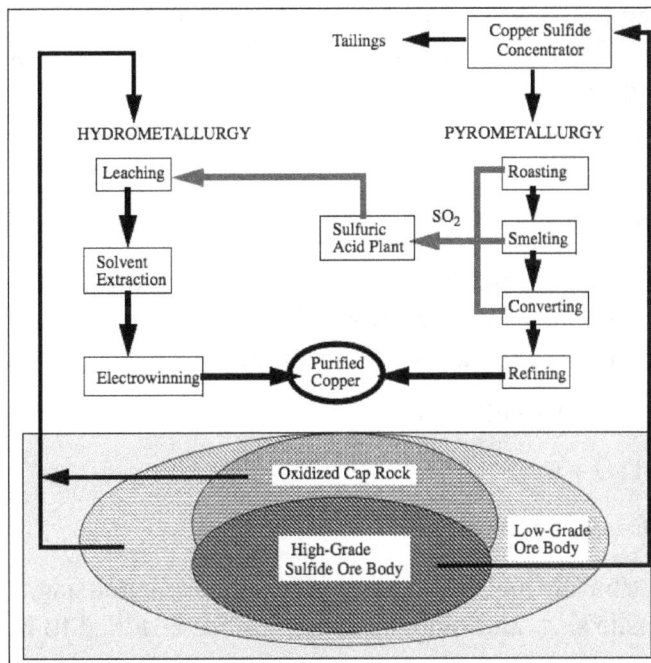

Processing of a typical copper oxide/copper sulfide ore body by a combination of pyrometallurgical and hydrometallurgical methods. The oxidized cap rock and the low-grade ore must be removed so that open-pit mining can reach the high-grade sulfides, and so they are readily available for hydrometallurgical processing.

Pyrometallurgy of Copper

The pyrometallurgical extraction of copper from sulfide minerals traditionally consists of the following basic steps:

- Roasting of sulfide concentrate.
- Matte smelting.

- Converting.
- Fire-refining.

After fire-refining, the copper is about 99.5% pure, and is further treated by electrolytic refining to produce extremely high purity copper.

There are several different processes for carrying out these operations, and in some cases several steps are carried out simultaneously.

Reactions

The basic chemical reactions in copper pyrometallurgy are as follows. Note: These reactions are grossly oversimplified, and are assuming a feed that contains no sulfides other than chalcopyrite. In a real process, there are many other oxides, sulfides, and sulfates involved besides the ones shown, and many side reactions and alternate reaction paths occur which greatly complicate the chemistry.

1. Roasting reactions: Exothermic reactions that provide heat, and that oxidize excess iron so that it can be removed during the smelting process along with silicates that are present in the ore. About one-third of the sulfide in the ore is oxidized in this step, producing a mixture of copper and iron sulfides, sulfates, and oxides:

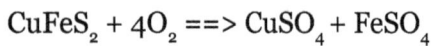

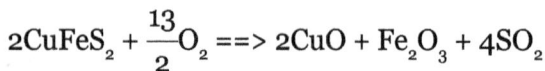

$$CuFeS_2 + 4O_2 ==> CuSO_4 + FeSO_4$$

$$2CuFeS_2 + \frac{13}{2}O_2 ==> 2CuO + Fe_2O_3 + 4SO_2$$

2. Smelting reactions: Any copper that was oxidized during roasting is re-reduced by part of the remaining iron sulfide so that the copper will not be lost in the smelter slag. The copper then forms a low-viscosity "matte" that melts and separates from the silicate slag.

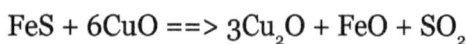

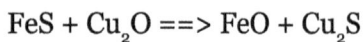

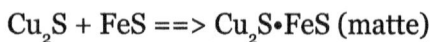

$$FeS + 6CuO ==> 3Cu_2O + FeO + SO_2$$

$$FeS + Cu_2O ==> FeO + Cu_2S$$

$$Cu_2S + FeS ==> Cu_2S \bullet FeS \text{ (matte)}$$

3. Converting reactions: After the matte is separated from the smelter slag, it is oxidized to produce sulfur dioxide, an iron oxide slag, and metallic copper. Silica is added to help form the iron oxide slag.

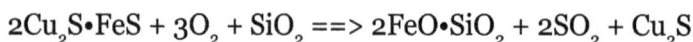

$$2Cu_2S \bullet FeS + 3O_2 + SiO_2 ==> 2FeO \bullet SiO_2 + 2SO_2 + Cu_2S$$

$$Cu_2S + O_2 ==> 2Cu + SO_2$$

4. Fire-refining reactions: The copper from the converter ("blister copper") contains a considerable amount of dissolved oxygen and copper oxide. The oxygen is removed by adding a hydrocarbon as a reducing agent.

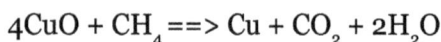

$$4CuO + CH_4 ==> Cu + CO_2 + 2H_2O$$

Roasting

Roasting consists of partially oxidizing the sulfide mineral concentrate with air, at between 500°C and 700°C. The main roasting reactions for chalcopyrite are:

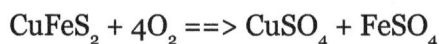

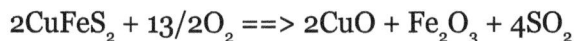

$$CuFeS_2 + 4O_2 ==> CuSO_4 + FeSO_4$$

$$2CuFeS_2 + 13/2O_2 ==> 2CuO + Fe_2O_3 + 4SO_2$$

It is important to note that only about one-third of the sulfides in the mineral concentrate are actually oxidized, with the rest remaining as sulfide minerals.

Objectives of roasting are to remove part of the sulfur, convert excess iron sulfides into oxides and sulfates that can be more easily removed during smelting, and to pre-heat the concentrate to reduce the amount of energy needed by the smelter.

The reactions that occur in smelting are all exothermic, and so roasting is an autogenous process that requires little or no additional fuel.

When roasting sulfides for smelting, the sulfide minerals are only partially oxidized. The objective is to convert the iron sulfides into iron oxide and iron sulfate, while keeping the bulk of the copper in the sulfide form.

During roasting, the gases produced are 5-15% SO_2, which is concentrated enough to be used for sulfuric acid production.

There are several different types of roaster in use. The most common types are:

- Hearth Roaster: This type of furnace is suitable for coarse particles, and is common in older facilities. The capacity is on the order of 100-200 tons/day.

- Flash or Suspension Roaster: These furnaces process very fine particles, and take advantage of the rapid reaction rate to provide autogenous heating. The capacity is approximately 3 to 4 times that of a similar-sized hearth roaster.

- Fluidized Bed Roaster: Gas is bubbled up through a bed of particles, with the particles large enough that they are not swept out of the furnace, but small enough that the gas can expand the bed so that it will behave as a fluid. These reactors provide excellent gas/solid contact, but tend to generate a lot of dust.

Smelting

The matte smelting process consists of melting the roasted concentrate to form two liquid phases: a sulfide "matte" which contains the copper, and an oxide slag which is insoluble in the matte, and contains iron oxides, silicates, and other impurities.

Smelting is carried out by melting the charge at about 1200°C, usually with a silica flux to make the slag more fluid. The silica, alumina, iron oxides, calcium oxides, and other minor oxides form.

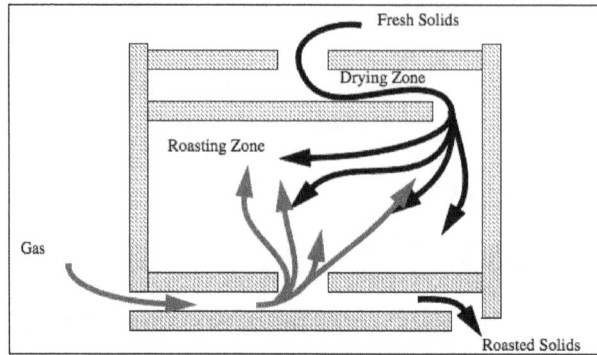

Flash Roasting. The heat produced by the roasting reaction is used to dry the incoming ore and bring it up to roasting temperature. If the reaction is sufficiently exothermic, the ore can even be melted.

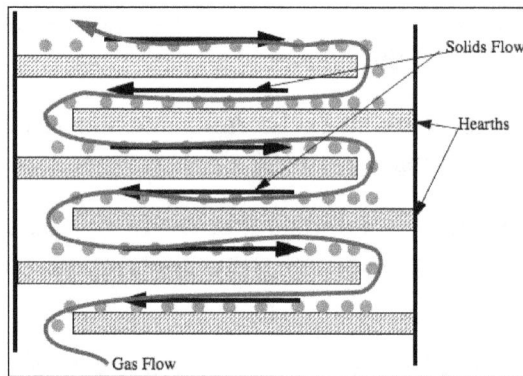

Hearth Roaster. Typically, these have 6 to 12 hearths, and most of the roasting occurs as the particles drop from one hearth to the next.

molten slag, while the copper, sulfur, unoxidized iron, and precious metals form the matte. The slag is lighter than the matte, and so it floats to the surface to be tapped off and disposed of.

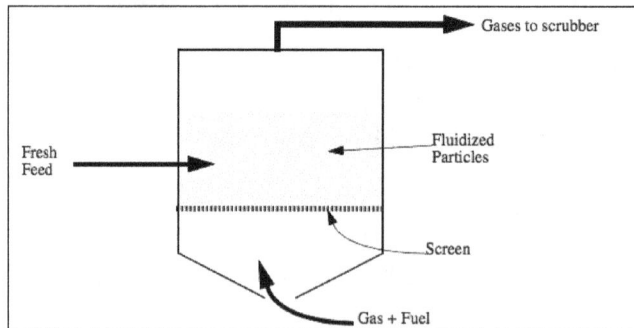

Fluidized Bed Reactor.

Relatively few chemical reactions are carried out during matte smelting, as its main purpose is simply to allow compounds to segregate into whichever phase they are most soluble in (slag or matte). The main important reaction is the conversion of copper oxides (that were formed during roasting) back into copper sulfide so that they will go into the matte phase:

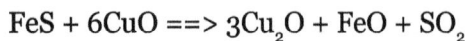

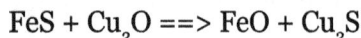

$$FeS + 6CuO ==> 3Cu_2O + FeO + SO_2$$

$$FeS + Cu_2O ==> FeO + Cu_2S$$

$$Cu_2S + FeS ==> Cu_2S \cdot xFeS \text{ (matte)}$$

In order for matte smelting to work, it is very important that the feed be only partially oxidized, and that enough sulfur remains in the charge for all of the copper to form copper sulfides. Matte smelting is carried out in a neutral or slightly reducing atmosphere to prevent overoxidation of the charge.

A typical matte consists of Cu_2S and FeS, and can be anywhere from 30% Cu to 80% Cu. At smelting temperatures, the viscosity of the matte is approximately 10 centipoise (cP), which is approximately 10 times more viscous than water.

Smelter slags typically have the following composition:

- Fe:(as FeO, $Fe3O_4$)   30 - 40%

- SiO_2:(from fluxes, or recycled slags)   35 - 40%

- Al_2O_3:   up to 10%

- CaO: up to 10%

The slag must have the following properties:

- Immiscible with the matte phase.

- Low solubility of Cu_2S in the slag.

- Good fluidity, to minimize entrainment of droplets of copper-bearing material into the slag.

In order to achieve these properties, the composition of the slag must be carefully controlled. It is particularly important that the viscosity be kept as low as possible. Slags are very viscous (500 - 2000 centipoise), which is many times higher than that of the matte phase, and high viscosity results in the slag entrapping more droplets of the matte.

Slag viscosity increases as it becomes more oxidized, because the iron becomes particles of solid magnetite which are not molten at copper smelting temperatures. It is therefore important to avoid overoxidizing the slag, so that the iron remains as liquid FeO.

Matte smelting can be carried out in a wide range of furnace types. Most commonly, it is done in reverberatory furnaces, although it can also be done in blast furnaces or electric furnaces. It can also be combined with roasting, as in the Outokumpu or INCO flash-smelting furnaces. In flashsmelting, the exothermic roasting reaction provides much of the energy needed to melt the matte and slag, and so the energy cost is reduced.

Converting

After the slag and matte are separated, the matte must be converted to metallic copper. This is done by selective oxidation of the matte with oxygen, to oxidize the sulfur but leave the copper as metal. This is a batch process, carried out in a horizontal cylindrical reactor called the Pierce-Smith Converter, which is shown in figure. This converter can be rotated into different positions for charging with matte, blowing with air, skimming off slag, or dumping finished copper. This is a batch reactor, rather than a continuous process.

Converting is carried out in two stages: an iron-removal stage and a copper-making stage.

Iron removal: For iron removal, silica is added as a flux to keep the slag molten, and air is blown into the converter to oxidize the iron sulfide, as shown in the following reaction:

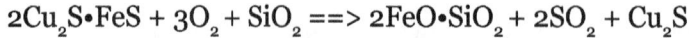

$$2Cu_2S{\cdot}FeS + 3O_2 + SiO_2 ==> 2FeO{\cdot}SiO_2 + 2SO_2 + Cu_2S$$

The oxidized iron and silica form a slag that is insoluble in the matte. Since iron oxidizes more readily than copper, this reaction continues until the matte contains less than 1% Fe.

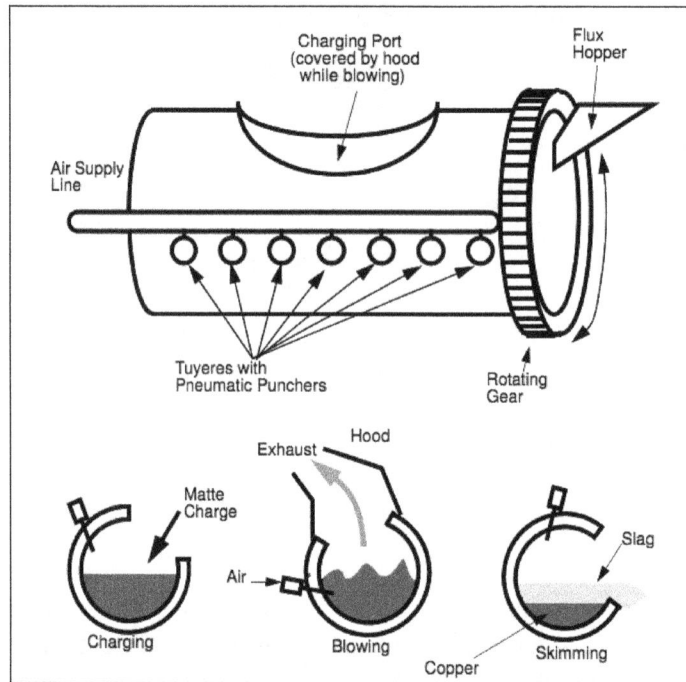

Schematic of a Pierce-Smith copper converter. During the converting process, air is injected through the tuyeres, and off-gases are collected by a removable hood that covers the charging port. The tuyeres are equipped with pneumatic "punchers" to break through any slag or other solids that may solidify over the tuyere opening.

The converter is then rotated to skim off the iron-bearing slag, and the slag is disposed of.

Once the slag has been skimmed, there is room for additional matte in the converter, and it is re-charged and the iron is oxidized and skimmed from the fresh matte.

This is continued until the converter is filled with Cu_2S, with very little remaining iron.

Because of the viscosity of the converter slag, it always contains about 2-10% copper. The slag is therefore generally returned to the processing plant to be crushed, ground, and treated along with the incoming ore, to recover the copper.

Copper making: Once the iron is removed, blowing is resumed. Since the iron is no longer present to consume the oxygen, the sulfur can be selectively oxidized to leave behind metallic copper, as shown in the following reaction:

$$Cu_2S + O_2 ==> 2Cu + SO_2$$

$$2Cu + O_2 ==> 2CuO \text{ (undesirable side reaction)}$$

Blowing is continued until the copper sulfide has been completely oxidized to metallic copper and sulfur dioxide. The resulting product is called "blister copper", and is approximately 98.5% pure copper. The name is due to the fact that, if it is allowed to solidify, this copper will contain "blisters" due to evolution of SO_2.

Fire Refining

Copper is fire refined by transferring it into a second converter, where it is first blown with air to completely oxidize any sulfur present. In the process of doing this, a significant amount of the copper also oxidizes, and a great deal of oxygen is dissolved in the molten copper as well. Traditionally, the oxygen was removed by a process known as "poling". This consists of adding a "pole", which is a green tree trunk (usually a softwood such as pine) to the copper. The pole is consumed, and the hydrocarbons it releases then reduce any copper oxide back to metallic copper. Poling was widely used in the past, but modern smelters typically inject gaseous hydrocarbons, which gives better control of the process.

In the past, fire-refined copper was the finished product, and was cast into ingots for sale. Today, almost all copper is electrorefined, and so the fire-refined copper is cast into electrolytic anodes.

Combined Processes

Each of the steps described in coppermaking (roasting, smelting, and converting) are controlled oxidation reactions with air. This suggests that, with proper design and control, it should be possible to combine individual steps to simplify the process, and ultimately to carry out the entire process in a single continuous process. The overall reaction for the entire continuous process can be written as:

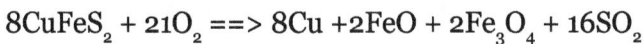

$$8CuFeS_2 + 21O_2 ==> 8Cu + 2FeO + 2Fe_3O_4 + 16SO_2$$

Since the late 1950s, processes have been under development that combine roasting and smelting into a single continuous process, and that combine continuous converting with smelting and roasting.

Several processes have been developed that accomplish this to varying degrees. These are the Outokumpu Flash-Smelting process, the Isasmelt process, the Noranda process, the Worcra process, and the Mitsubishi process. All of these processes are generally similar to each other.

Flash Smelting

The Outokumpu flash smelting process combines a flash roast with a smelting bath, as shown in figure. In this topic, sulfide concentrate is injected through a burner along with preheated air, and partially burned as it falls through a flash-roasting zone.

Sulfide helps to melt the concentrate, with supplemental fuel injected into the flash-roasting area to provide the remaining heat.

By the time the ore reaches the bottom of the smelter, it has been melted. It then segregates into a slag phase and a matte phase, which are tapped off separately.

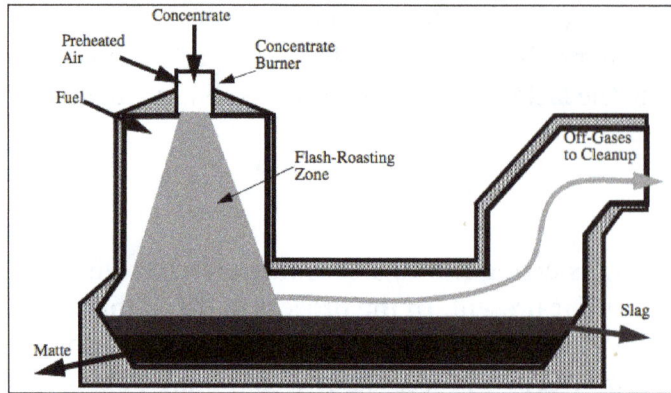

Outokumpu flash smelting process. Incoming ore concentrate is entrained in
the incoming air, and partially combusted to simultaneously roast and melt the concentrate.

This technology has been available for nearly 50 years, and is widely used. In addition to the flash smelting, Outokumpu and Kennecott have developed a convertor system, known as "flash converting", which works on similar principles.

Isasmelt Process

The Isasmelt process was developed at Mount Isa Mines, originally to smelt lead ores. The heart of the process is a top-blown vessel, as shown in figure. Concentrate is pelletized along with supplemental fuel (coal), fluxes, and any other necessary additives. The pellets are then fed to the vessel, and air that has been enriched to about 45% oxygen is blown into the top using a lance. Roasting and fuel combustion occur in the molten bath, and the molten material is sent to a holding furnace where the slag and matte are separated.

Work is currently underway to adapt the Isasmelt furnace for use as a converter, so that the entire process could be run continuously.

Noranda Process

The Noranda process uses a cylindrical reactor similar to the Pierce-Smith converter. Pelletized feed is added along with coal fines at the feed-end burner, where it is roasted, melted, and forms into slag and matte phases. The slag is tapped off, while air is injected into the tuyeres to continuously produce metallic copper.

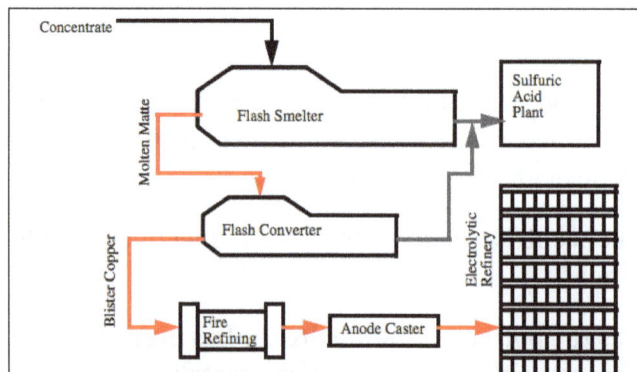

Integrated flash smelting/flash converting copper production facility.

Worcra Process

The Worcra process was developed at the same time as the Noranda process, and is similar in most respects. The main difference is that it is a stationary furnace rather than a tiltable vessel, and the air is injected from the top through air lances, rather than through submerged tuyeres as is the case in the Noranda vessel. It also includes a settling basin at the slag discharge end, to remove droplets of matte that were entrained into the slag.

Mitsubishi Process

The Mitsubishi process uses three furnaces in series rather than a single furnace. The first unit is essentially a flash-smelting unit, used to simultaneously roast and melt the ore into a slag and matte. The molten material is then sent to an electric furnace which keeps it molten using electric arc heating under a reducing atmosphere, to allow the slag and matte to separate completely. The matte is then run to a continuous converter to be converted into metal. The use of three furnaces instead of one gives better control of the process than in either the Noranda or Worcra methods, and the design of the furnaces to provide continuous flow rather than batch flow gives a higher throughput than is the case with the traditional methods.

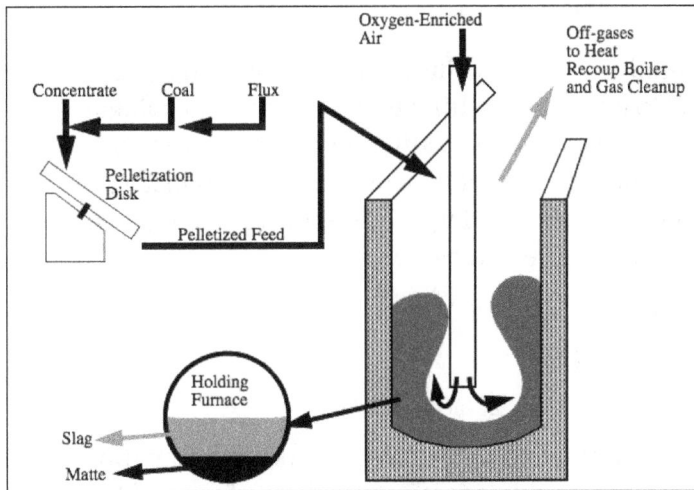

Isasmelt smelting vessel.

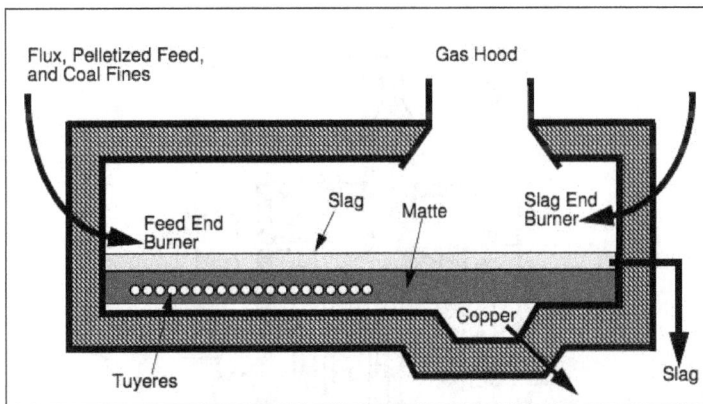

Schematic of the Noranda single-step reactor. Feed at the feed end first melts and segregates into slag and matte. The slag is then tapped off, while the matte is converted to copper.

ELECTROMETALLURGY

Electrometallurgy is an important branch of metallurgy which uses electrical energy in winning of metals from leach liquors and refining of crude metals obtained via the pyro-metallurgical route as well as by hydrometallurgical and electrometallurgical routes. This branch of metallurgy also includes the production of metals and alloys by electrothermic smelting.

Electrometallurgy is a common extraction process for the more reactive metals, e.g., for aluminum and metals above it in the electrochemical series. It is one method of extracting copper and in the purification of copper. During electrolysis, electrons are being added directly to the metal ions at the cathode (the negative electrode). The downside (particularly in the aluminum case) is the cost of the electricity. An advantage is that it can produce very pure metals.

Preparation of Sodium

The most important method for the production of sodium is the electrolysis of molten sodium chloride; the set-up is a Downs cell, shown in figure. The reaction involved in this process is:

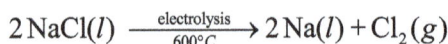

$$2\,NaCl(l) \xrightarrow[600°C]{electrolysis} 2\,Na(l) + Cl_2(g)$$

The electrolysis cell contains molten sodium chloride (melting point 801 °C), to which calcium chloride has been added to lower the melting point to 600 °C (a colligative effect). The passage of a direct current through the cell causes the sodium ions to migrate to the negatively charged cathode and pick up electrons, reducing the ions to sodium metal. Chloride ions migrate to the positively charged anode, lose electrons, and undergo oxidation to chlorine gas. The overall cell reaction comes from adding the following reactions:

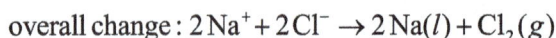

at the cathode : $2\,Na^+ + 2e^- \rightarrow 2\,Na(l)$

at the anode : $2\,Cl^- \rightarrow Cl_2(g) + 2e^-$

overall change : $2\,Na^+ + 2\,Cl^- \rightarrow 2\,Na(l) + Cl_2(g)$

Separation of the molten sodium and chlorine prevents recombination. The liquid sodium, which is less dense than molten sodium chloride, floats to the surface and flows into a collector. The gaseous chlorine goes to storage tanks. Chlorine is also a valuable product.

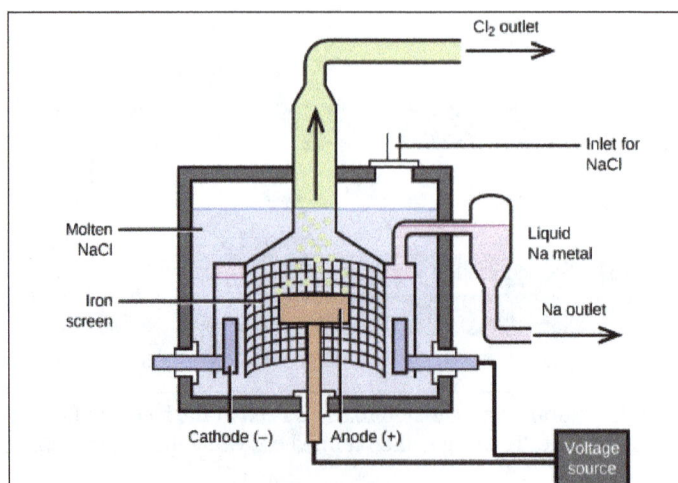

Pure sodium metal is isolated by electrolysis of molten sodium chloride using a Downs cell. It is not possible to isolate sodium by electrolysis of aqueous solutions of sodium salts because hydrogen ions are more easily reduced than are sodium ions; as a result, hydrogen gas forms at the cathode instead of the desired sodium metal. The high temperature required to melt NaCl means that liquid sodium metal forms.

Preparation of Aluminum

The preparation of aluminum utilizes a process invented in 1886 by Charles M. Hall, who began to work on the problem while a student at Oberlin College in Ohio. Paul L. T. Héroult discovered the process independently a month or two later in France. In honor to the two inventors, this electrolysis cell is known as the Hall–Héroult cell. The Hall–Héroult cell is an electrolysis cell for the production of aluminum.

The production of aluminum begins with the purification of bauxite, the most common source of aluminum. The reaction of bauxite, AlO(OH), with hot sodium hydroxide forms soluble sodium aluminate, while clay and other impurities remain undissolved:

$$AlO(OH)(s) + NaOH(aq) + H_2O(l) \rightarrow Na[Al(OH)_4](aq)$$

After the removal of the impurities by filtration, the addition of acid to the aluminate leads to the reprecipitation of aluminum hydroxide:

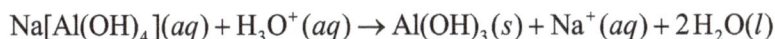

$$Na[Al(OH)_4](aq) + H_3O^+(aq) \rightarrow Al(OH)_3(s) + Na^+(aq) + 2H_2O(l)$$

The next step is to remove the precipitated aluminum hydroxide by filtration. Heating the hydroxide produces aluminum oxide, Al_2O_3, which dissolves in a molten mixture of cryolite, Na_3AlF_6, and calcium fluoride, CaF_2. Electrolysis of this solution takes place in a cell like that. Reduction of aluminum ions to the metal occurs at the cathode, while oxygen, carbon monoxide, and carbon dioxide form at the anode.

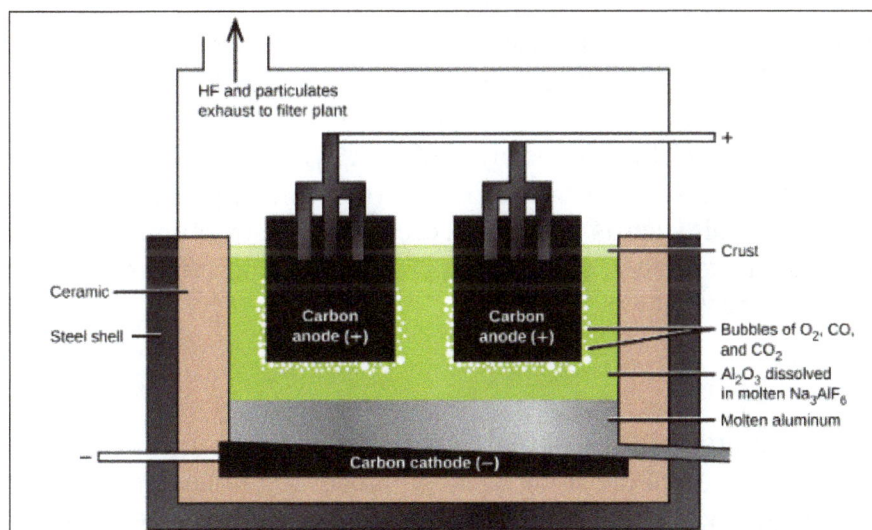

An electrolytic cell is used for the production of aluminum. The electrolysis of a solution of cryolite and calcium fluoride results in aluminum metal at the cathode, and oxygen, carbon monoxide, and carbon dioxide at the anode.

Preparation of Magnesium

Magnesium is the other metal that is isolated in large quantities by electrolysis. Seawater, which contains approximately 0.5% magnesium chloride, serves as the major source of magnesium. Addition of calcium hydroxide to seawater precipitates magnesium hydroxide. The addition of hydrochloric acid to magnesium hydroxide, followed by evaporation of the resultant aqueous solution, leaves pure magnesium chloride. The electrolysis of molten magnesium chloride forms liquid magnesium and chlorine gas:

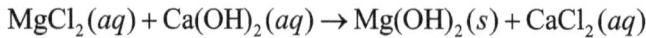

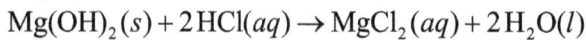

$$MgCl_2(aq) + Ca(OH)_2(aq) \rightarrow Mg(OH)_2(s) + CaCl_2(aq)$$
$$Mg(OH)_2(s) + 2HCl(aq) \rightarrow MgCl_2(aq) + 2H_2O(l)$$
$$MgCl_2(l) \rightarrow Mg(l) + Cl_2(g)$$

Some production facilities have moved away from electrolysis completely.

Electrowinning

Electrowinning, also called electroextraction, is the electrodeposition of metals from their ores that have been put in solution via a process commonly referred to as leaching. Electrorefining uses a similar process to remove impurities from a metal. Both processes use electroplating on a large scale and are important techniques for the economical and straightforward purification of non-ferrous metals. The resulting metals are said to be *electrowon*.

In electrowinning, a current is passed from an inert anode through a liquid *leach* solution containing the metal so that the metal is extracted as it is deposited in an electroplating process onto the cathode. In electrorefining, the anodes consist of unrefined impure metal, and as the current passes through the acidic electrolyte the anodes are corroded into the solution so that the electroplating process deposits refined pure metal onto the cathodes.

Applications

The most common electrowon metals are lead, copper, gold, silver, zinc, aluminium, chromium, cobalt, manganese, and the rare-earth and alkali metals. For aluminium, this is the only production process employed. Several industrially important active metals (which react strongly with water) are produced commercially by electrolysis of their pyrochemical molten salts. Experiments using electrorefining to process spent nuclear fuel have been carried out. Electrorefining may be able to separate heavy metals such as plutonium, caesium, and strontium from the less-toxic bulk of uranium. Many electroextraction systems are also available to remove toxic (and sometimes valuable) metals from industrial waste streams.

Process

Most metals occur in nature in their oxidized form (ores) and thus must be reduced to their metallic forms. The ore is dissolved following some preprocessing in an aqueous electrolyte or in a molten salt and the resulting solution is electrolyzed. The metal is deposited on the cathode (either in solid or in liquid form), while the anodic reaction is usually oxygen evolution. Several metals are naturally present as metal sulfides; these include copper, lead, molybdenum, cadmium, nickel, silver, cobalt, and zinc. In addition, gold and platinum group metals are associated with sulfidic base metal ores. Most metal sulfides or their salts, are electrically conductive and this allows electrochemical redox reactions to efficiently occur in the molten state or in aqueous solutions.

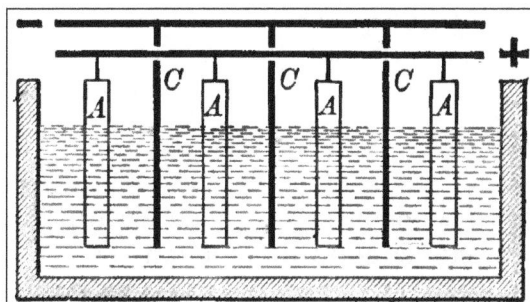

Apparatus for electrolytic refining of copper.

Some metals, such as nickel do not electrolyze out but remain in the electrolyte solution. These are then reduced by chemical reactions to refine the metal. Other metals, which during the processing of the target metal have been reduced but not deposited at the cathode, sink to the bottom of the electrolytic cell, where they form a substance referred to as *anode sludge* or *anode slime*. The metals in this sludge can be removed by standard pyrorefining methods.

Because metal deposition rates are related to available surface area, maintaining properly working cathodes is important. Two cathode types exist, flat-plate and reticulated cathodes, each with its own advantages. Flat-plate cathodes can be cleaned and reused, and plated metals recovered. Reticulated cathodes have a much higher deposition rate compared to flat-plate cathodes. However, they are not reusable and must be sent off for recycling. Alternatively, starter cathodes of pre-refined metal can be used, which become an integral part of the finished metal ready for rolling or further processing.

Electro Refining

Electro refining is one of a collection of electrochemical processes which are primarily concerned with the extraction of metals from their ores and or the subsequent refining of the metals to high purity. The main advantages of electro refining processes are they are designed to handle a wide variation in the quality of the base scrap and conversely can provide a particularly high purity of end product material.

Electrochemical processing is used both in the primary extraction of metals from their ores and in the subsequent refining of metals to high purity. Both operations are accomplished in an electrolytic cell, a device that permits electrical energy to perform chemical work. This occurs by the transfer of electrical charge between two electrodes immersed in an ionically conducting liquid (electrolyte) containing metal dissolved as positive ions.

At the negatively charged cathode the metal cations acquire electrons (are reduced), and deposit as neutral metal atoms. At the positively charged anode there are two possible reactions, depending upon the type of cell. In an electrowinning cell the dissolution of the anode metal itself occurs. The more noble metals such as copper and zinc are electrolyzed from aqueous electrolytes, whereas reactive metals such as aluminum and magnesium are electrolyzed from electrolytes of their fused salts.

In an electrorefining process, the anode is the impure metal and the impurities must be lost during the passage of the metal from the anode to the cathode during electrolysis, i.e. the electrode reactions are, at the anode:

$$M \rightarrow M^{n+} + ne^-$$

and at the cathode:

$$M^{n+} + ne^- \rightarrow M$$

Electrorefining is a much more common process than electrowinning and such plants occur throughout the world on scales between 1000-100,000 ton/year.

Usually they are part of a larger operation to separate and recover pure metals from both scrap and primary ores. Therefore, the process must be designed to handle a variable-quality metal feed and lead to a concentration of all the metals present in a form which can be treated further. Electrorefining often provides a particularly high purity of metal.

Electrorefining processes using a molten salt or non-aqueous electrolyte are used and, indeed, are the subject of further development. This is due to the possibilities they offer for increasing current densities and refining via lower oxidation states not stable in water (e.g. refining of copper via Cu^+ would almost halve the energy requirement). However, aqueous processes presently predominate due to their ease of handling, more developed chemistry and familiarity with aqueous process liquors and electrolytes.

Aqueous electrorefining: The conditions used for the refining of five metals are summarized in table. The electrolyte and other conditions must be selected so that both the anodic dissolution and the deposition of the metal occur with high efficiency while none of the impurity metals can transfer from the anode to the cathode. Certainly there must be no passivation of the anode and the objective is to obtain a good-quality, often highly crystalline, deposit at the cathode.

Where necessary, additives are added to the electrolyte to enforce the correct behavior at both electrodes. Chloride ion is a common addition to enhance the dissolution process and, where essential, organic additives are used to modify the cathode deposit. Since, however, organic compounds can be occluded to some extent and reduce the purity of the metal, their use is avoided when possible.

Table: The conditions used for the refining of five metals.

Metal	Concentration of components in electrolyte/g dm^{-3}	I/mA cm^{-2}	$-$Cell voltage/V	T/°C	Current efficiency/%	Impurity metals	
						Slime	Solution
Cu	$CuSO_4$ (100–140) H_2SO_4 (180–250)	10–20	0.15–0.30	60	95	Ag, Au, Ni, Pb, Sb	Ni, As, Fe, Co
Ni	$NiSO_4$ (140–160) NaCl (90) H_3BO_3 (10–20)	15–20	1.5–3.0	60	98	Ag, Au, Pt	Cu, Co
Co	$CoSO_4$ (150–160) Na_2SO_4 (120–140) NaCl (15–20) H_3BO_3 (10–20)	15–20	1.5–3.0	60	75–85	—	Ni, Cu
Pb	Pb^{2+} (60–80) H_2SiF_6 (50–100)	15–25	0.3–0.6	30–50	95	Bi, Ag, Au, Sb	—
Sn	Na_2SnO_3 (40–80) NaOH (8–20)	5–15	0.3–0.6	20–60	65	Pb, Sb	—

Electrorefining is widely used for the purification and production of copper that is suitable for electrical applications. Such plants exist throughout the world on production scales between 1000 and 100 000 t/a.

In an electrorefining process, the anode is the impure metal and the impurities are lost during the passage of metal from the anode to the cathode during electrolysis. The electrode reactions in the case of Cu electrorefining are as follows:

Anode reaction: $Cu \rightarrow Cu^{2+} + 2e^-$

If Ni and Fe are also present in the impure anode, they will dissolve as follows:

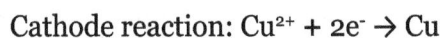

$Ni \rightarrow Ni^{2+} + 2e^-$

$Fe \rightarrow Fe^{2+} + 2e^-$

Cathode reaction: $Cu^{2+} + 2e^- \rightarrow Cu$

A suitable voltage is applied to the electrodes to cause oxidation of copper metal at the anode and reduction of Cu^{2+} to form copper metal at the cathode. This works efficiently because copper is both oxidized and reduced more readily than water. Metallic impurities with a lower reduction potential than copper are less noble and will readily dissolve at the anode but do not plate at the cathode. More noble metals with a higher reduction potential are not dissolved at the anode, instead they collect at the bottom of the cell as anode slimes. The anode slimes can be captured and processed to recover the valuable metals.

Cell voltage and current density are the two important parameters in copper electrorefining. The total voltage is determined by the equilibrium cell voltage, anodic and cathodic overpotential and Ohmic potential drop in electrolyte, hardware and power supply. It is generally accepted that copper production increases with an increase of current density at the cost of current efficiency. The electrolyte that serves as a carrier for the Cu^{2+} ions is sometimes in the form of a molten salt or non-aqueous electrolyte.

These forms of electrolyte offer opportunities for increasing current densities and refining via lower oxidation states that are not stable in water. However, aqueous processes are the most widely used due to the ease of handling, more developed chemistry and familiarity with aqueous process liquors and electrolytes. Cu electrorefining is typically conducted using a sulphate medium for the transport of Cu ions.

In the study of T. Takenaka et al. the electrorefining of Mg has been investigated in a molten salt system, and the electrolysis conditions for the effective purification have been discussed. A purified mixture of $MgCl_2$–NaCl–$CaCl_2$ was used as an electrolytic bath. Magnesium metal was dissolved anodically by potentiostatic electrolysis, and purified Mg was electrodeposited at the cathode.

A certain degree of cathodic overpotential was required for the effective electrodeposition of Mg metal, while large anodic overpotential directly caused the deterioration in the purity of Mg electrodeposit; it was necessary for the anodic overpotential to be less than 1.0V for good purification. Under the suitable electrolysis condition, the Fe content in the Mg electrodeposit was less than 10 ppm.

A couple of subjects on recycling Mg metal and its alloys have been also studied: purification of Mg alloy by an electrorefining technique and distinction of Mg alloys. It was shown that pure Mg metal was electrodeposited at the cathode by the electrorefining of Mg alloy. X-ray fluorescence analysis was applied to distinction of Mg alloys, and the measuring conditions were discussed. It was concluded that the electrorefining process and X-ray fluorescence analysis were usable for the recycling of Mg metal and alloys.

Electroplating

Electroplating involves passing an electric current through a solution called an electrolyte. This is done by dipping two terminals called electrodes into the electrolyte and connecting them into a circuit with a battery or other power supply. The electrodes and electrolyte are made from carefully chosen elements or compounds. When the electricity flows through the circuit they make, the electrolyte splits up and some of the metal atoms it contains are deposited in a thin layer on top of one of the electrodes—it becomes electroplated. All kinds of metals can be plated in this way, including gold, silver, tin, zinc, copper, cadmium, chromium, nickel, platinum, and lead.

Electroplating is very similar to electrolysis (using electricity to split up a chemical solution), which is the reverse of the process by which batteries produce electric currents. All these things are examples of electrochemistry: chemical reactions caused by or producing electricity that give scientifically or industrially useful end-products.

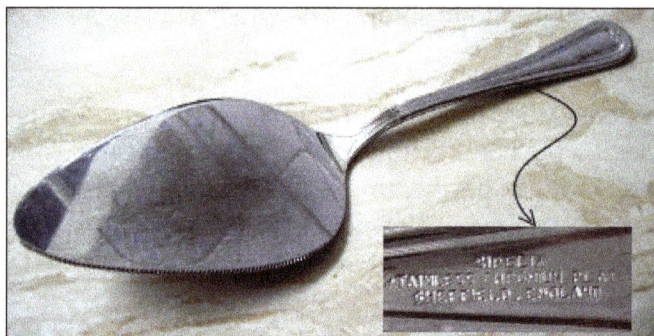

Silver cutlery is expensive and tarnishes; stainless steel plated with chromium is a good substitute for many people. Although it's rustproof and durable, the plating does eventually wear off, as you can see in the brownish area in the center of this pie server. "EPNS" marked on cutlery is a definitive sign of plating: it stands for electroplated nickel silver.

How does Electroplating Work?

First, you have to choose the right electrodes and electrolyte by figuring out the chemical reaction or reactions you want to happen when the electric current is switched on. The metal atoms that plate your object come from out of the electrolyte, so if you want to copper plate something you need an electrolyte made from a solution of a copper salt, while for gold plating you need a gold-based electrolyte—and so on.

Next, you have to ensure the electrode you want to plate is completely clean. Otherwise, when metal atoms from the electrolyte are deposited onto it, they won't form a good bond and they may simply rub off again. Generally, cleaning is done by dipping the electrode into a strong acid or alkaline solution or by (briefly) connecting the electroplating circuit in reverse. If the electrode is really clean, atoms from the plating metal bond to it effectively by joining very strongly onto the outside edges of its crystalline structure.

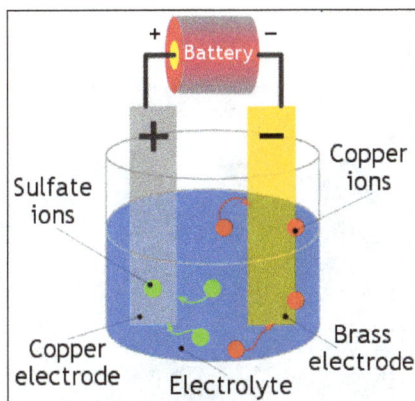

Copper-plating brass: You need a copper electrode (gray, left), a brass electrode (yellow, right), and some copper sulfate solution (blue). The brass electrode becomes negatively charged and attracts positively charged copper ions from the solution, which cling to it and form an outer coating of copper plate.

Now we're ready for the main part of electroplating. We need two electrodes made from different conducting materials, an electrolyte, and an electricity supply. Generally, one of the electrodes is made from the metal we're trying to plate and the electrolyte is a solution of a salt of the same metal. So, for example, if we're copper plating some brass, we need a copper electrode, a brass electrode, and a solution of a copper-based compound such as copper sulfate solution. Metals such as gold and silver don't easily dissolve so have to be made into solutions using strong and dangerously unpleasant cyanide-based chemicals. The electrode that will be plated is generally made from a cheaper metal or a nonmetal coated with a conducting material such as graphite. Either way, it has to conduct electricity or no electric current will flow and no plating will occur.

We dip the two electrodes into the solution and connect them up into a circuit so the copper becomes the positive electrode (or anode) and the brass becomes the negative electrode (or cathode). When we switch on the power, the copper sulfate solution splits into ions (atoms with too few or too many electrons). Copper ions (which are positively charged) are attracted to the negatively charged brass electrode and slowly deposit on it—producing a thin later of copper plate. Meanwhile, sulfate ions (which are negatively charged) arrive at the positively charged copper anode, releasing electrons that move through the battery toward the negative, brass electrode.

It takes time for electroplated atoms to build up on the surface of the negative electrode. How long exactly depends on the strength of the electric current you use and the concentration of the electrolyte. Increasing either of these increases the speed at which ions and electrons move through the circuit and the speed of the plating process. As long as ions and electrons keep moving, current keeps flowing and the plating process continues.

Electroplating Plastics

The below figures' shows: Plated plastic is often used on parts that need the shiny finish of a metal without its strength or heaviness, and here are three examples from my own home. Top: The switch, hands, and bezel (dial surround) of this alarm clock look shiny and metallic, but they're actually plastic. Middle: Plumbing parts that don't need to be strong are often made from plated plastic so they stay cool to the touch and blend in with metal pipes. The temperature control on this shower (right, with the red button) is made of plastic, but looks similar to the main metal components on the left. Bottom: This USB computer microphone has been given a shiny plated finish to make it look expensive and high-quality.

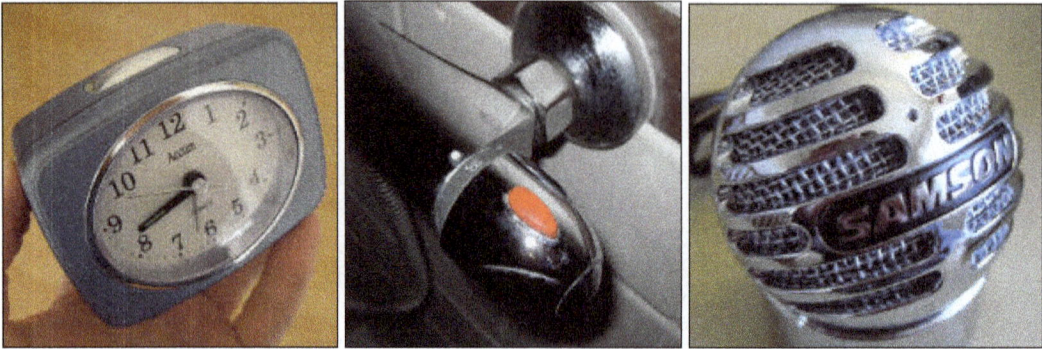

Inexpensive, easy to form into different shapes, lightweight, and disposable, plastics rapidly became the world's most commonplace and flexible materials in the 20th century. But, to many people, that's as much of a drawback as a benefit: plastics are cheap and cheerful—and that's exactly what they look like. One solution is to coat a cheap plastic with a thin layer of metal to give it all the benefits of plastic with the attractive, shiny finish of metal. Many different plastics can be plated this way, including ABS, phenolic plastics, urea-formaldehyde, nylon, and polycarbonate. You'll often find parts on cars, plumbing, household, and electrical fittings that look metallic but are, in fact, plated plastic. They're lighter, cheaper, rustproof, and don't require any polishing after plating.

How are Plastics Electroplated?

Plastics generally don't conduct electricity. In theory, that should completely rule out electroplating; in practice, it simply means we have to give our plastic an extra treatment to make it electrically conducting before we start. There are several different steps involved. First, the plastic has to be scrupulously cleaned to remove things like dust, dirt, grease, and surface marks. Next, it's etched with acid and treated with a catalyst (a chemical reaction accelerator) to make sure that a coating will stick to its surface. Then it's dipped in a bath of copper or nickel (copper is more common) to give it a very thin coating of electrically conducting metal (less than a micron, 1μm, or one thousandth of a millimeter thick). Once that's done, it can be electroplated just like a metal. Depending on how much wear and tear the plated part has to withstand, the coating can be anything from about 10–30 microns thick.

Why use Electroplating?

Electroplating is generally done for two quite different reasons. Metals such as gold and silver are plated for decoration: it's cheaper to have gold- or silver-plated jewelry than solid items made from these heavy, expensive, precious substances. Metals such as tin and zinc (which aren't especially attractive to look at) are plated to give them a protective outer later. For example, food containers are often tin plated to make them resistant to corrosion, while many everyday items made from iron are plated with zinc (in a process called galvanization) for the same reason. Some forms of electroplating are both protective and decorative. Car fenders and "trim," for example, were once widely made from tough steel plated with chromium to make them both attractively shiny and rust-resistant (inexpensive and naturally rustproof plastics are now more likely to be used on cars instead). Alloys such as brass and bronze can be plated too, by arranging for the electrolyte to contain salts of all the metals that need to be present in the alloy. Electroplating is also used for making duplicates of printing plates in a process called electrotyping and for electroforming (an alternative to casting objects from molten metals).

This car wheel is made from aluminum *metal plated with* nickel *in a more environmentally friendly process developed by Metal Arts Company, Inc. The* Microsmooth process uses about 30 percent less electricity, nearly 60 percent less natural gas, and half the water that conventional plating processes need.

Electroforming

Electroforming is a metal forming process that forms parts through electrodeposition on a model, known in the industry as a mandrel. Conductive (metallic) mandrels are passivated (chemically) to preclude 'plating' and thereby to allow subsequent separation of the finished electroform. Non-conductive (glass, silicon, plastic) mandrels require the deposition of a conductive layer prior to electrodeposition. Conductive layers can be deposited chemically, or using vacuum deposition techniques (e.g., gold sputtering). The outer surface of the mandrel forms the inner surface of the form.

The process involves high current through very clean water, having no more than about 5 parts per million organic contamination. The 'thrown' ions find their missing electrons on the mandrel, which is in electrical contact with the cathode of the electroforming tank. The ions deposit as neutral metal atoms, which bind to each other. Metal is electrodeposited until it is strong enough to be self-supporting. The mandrel is most often separated intact or dissolved away after forming, but occasionally (as in the case in decorative electroforming) left in place.

The surface of the finished part that was in intimate contact with the mandrel is rendered in fine detail with respect to the original, and is not subject to the shrinkage that would normally be experienced in a foundry cast metal object, or the tool marks of a milled part. The side of the part that was in contact with the electroforming solution is less well defined, and that loss of definition increases with thickness of the deposit. In extreme cases, where a thickness of several millimetres is required, there is preferential build-up of material on sharp outside edges and corners. This tendency can be reduced by a process known as periodic reverse, where the electroforming current is reversed for short periods and the excess is preferentially etched away. The finished form can either be the finished part, or can be used in a subsequent process to produce a positive of the original mandrel shape, such as with vinyl record-stamper manufacture.

In recent years, due to its ability to replicate a mandrel surface very precisely with practically no loss of fidelity, electroforming has taken on new importance in the fabrication of micro and nano-scale metallic devices and in producing precision injection moulds with micro- and nano-scale features for production of non-metallic micro-moulded objects.

Process

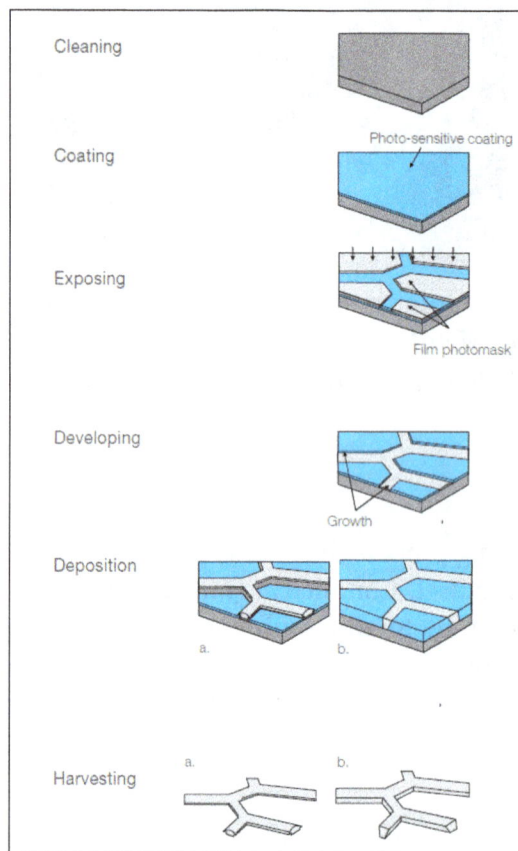

Electroforming process detail.

In the basic electroforming process, an electrolytic bath is used to deposit nickel or other electroformable metals onto a conductive patterned surface, such as stainless steel. Once the deposited material has been built up to the desired thickness, the master electroform is parted from the substrate. This process allows high-quality duplication of the mandrel and, therefore, permits quality production — at low unit costs with high repeatability and excellent process control.

If the mandrel is made of a non-conductive material it can be covered with a conductive coating.

The object being electroformed can be a permanent part of the end product or can be temporary (as in the case of wax), and removed later, leaving only the metal form, the "electroform". New technologies have made it possible for mandrels to be very complex. In order to facilitate the removal of the electroform from the mandrel, a mandrel is often made of aluminum. Because aluminum can easily be chemically dissolved, a complex electroform can be produced with near exactness.

Advantages and Disadvantages

The main advantage of electroforming is that it reproduces the external shape of the mandrel within one micrometer. Generally, forming an internal cavity accurately is more difficult than forming an external shape, however the opposite holds true for electroforming because the mandrel's exterior can be accurately machined.

Compared to other basic metal forming processes (casting, forging, stamping, deep drawing, machining and fabricating) electroforming is very effective when requirements call for extreme tolerances, complexity or light weight. The precision and resolution inherent in the photographically produced conductive patterned substrate, allows finer geometries to be produced to tighter tolerances while maintaining superior edge definition with a near optical finish. Electroformed metal is extremely pure, with superior properties over wrought metal due to its refined crystal structure. Multiple layers of electroformed metal can be molecularly bonded together, or to different substrate materials to produce complex structures with "grown-on" flanges and bosses.

Tolerances of 1.5 to 3 nanometres have been reported.

A wide variety of shapes and sizes can be made by electroforming, the principal limitation being the need to part the product from the mandrel. Since the fabrication of a product requires only a single pattern or mandrel, low production quantities can be made economically.

References

- Extractive-metallurgy, metallurgy, science: britannica.com, Retrieved 5 May, 2019

- Sagona, A.G.; Zimansky, P.E. (2009). Ancient Turkey. Routledge. ISBN 9780415481236. Archived from the original on 6 March 2016. Retrieved 26 August 2015

- Mineral-processing, technology: britannica.com, Retrieved 6 June, 2019

- "Releases/2007/04/070423100437". Sciencedaily.com. Archived from the original on 9 September 2015. Retrieved 26 August 2015

- Hydrometallurgy, technology: britannica.com, Retrieved 7 July, 2019

- United States Congress Office of Technology Assessment (1988). Copper, Technology & Competitiveness. DIANE Publishing. Pp. 142–143. ISBN 9781428922457

- Primary-Metals, kawatra: chem.mtu.edu, Retrieved 8 August, 2019

- "Advisory on EPA's Draft Technical Report entitled Considerations Related to Post- Closure Monitoring of Uranium In-Situ Leach/In-Situ Recovery (ISL/ISR) Sites". Retrieved 2012-10-13

- Leaching-in-metallurgy-and-metal-recovery: emew.com, Retrieved 9 January, 2019

- Alexander Watt, Electro-Deposition a Practical Treatise, Read Books (2008), p. 395. ISBN 1-4437-6683-6

Physical Metallurgy

Physical metallurgy is concerned with the physical properties of metals and alloys. Some of the properties which are studied within this field are mechanical, magnetic and thermal properties. This chapter has been carefully written to provide an easy understanding of the various properties and applications of physical metallurgy.

Physical metallurgy is the science of making useful products out of metals. Metal parts can be made in a variety of ways, depending on the shape, properties, and cost desired in the finished product. The desired properties may be electrical, mechanical, magnetic, or chemical in nature; all of them can be enhanced by alloying and heat treatment. The cost of a finished part is often determined more by its ease of manufacture than by the cost of the material. This has led to a wide variety of ways to form metals and to an active competition among different forming methods, as well as among different materials. Large parts may be made by casting. Thin products such as automobile fenders are made by forming metal sheets, while small parts are often made by powder metallurgy (pressing powder into a die and sintering it). Usually a metal part has the same properties throughout. However, if only the surface needs to be hard or corrosion-resistant, the desired performance can be obtained through a treatment that changes only the composition and strength of the surface.

Metallic Crystal Structures

Metals are used in engineering structures (e.g., automobiles, bridges, pressure vessels) because, in contrast to glass or ceramic, they can undergo appreciable plastic deformation before breaking. This plasticity stems from the simplicity of the arrangement of atoms in the crystals making up a piece of metal and the nondirectional nature of the bond between the atoms. Atoms can be arranged in many different ways in crystalline solids, but in metals the packing is in one of three simple forms. In the most ductile metals, atoms are arranged in a close-packed manner. If atoms were visualized as identical spheres and if these spheres were packed into planes in the closest possible manner, there would be two ways to stack close-packed planes one above another. One would lead to a crystal with hexagonal symmetry (called hexagonal close-packed, or hcp); the other would lead to a crystal with cubic symmetry that could also be visualized as an assembly of cubes with atoms at the corners and at the centre of each face (known as face-centred cubic, or fcc). Examples of metals with the hcp type of structure are magnesium, cadmium, zinc, and alpha titanium. Metals with the fcc structure include aluminum, copper, nickel, gamma iron, gold, and silver.

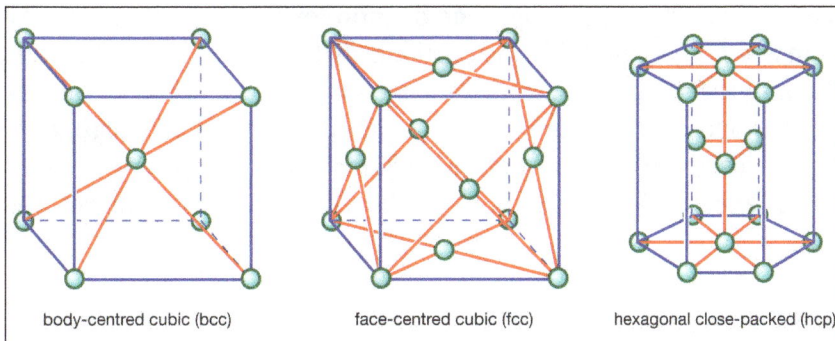

| body-centred cubic (bcc) | face-centred cubic (fcc) | hexagonal close-packed (hcp) |

The commonest metallic crystal structures.

The third common crystal structure in metals can be visualized as an assembly of cubes with atoms at the corners and an atom in the centre of each cube; this is known as body-centred cubic, or bcc. Examples of metals with the bcc structure are alpha iron, tungsten, chromium, and beta titanium.

Some metals, such as titanium and iron, exhibit different crystal structures at different temperatures. The lowest-temperature structure is labeled alpha (α), and higher-temperature structures beta (β), gamma (γ), and delta (δ). This allotropy, or transformation from one structure to another with changing temperature, leads to the marked changes in properties that can come from heat treatment.

When a metal undergoes a phase change from liquid to solid or from one crystal structure to another, the transformation begins with the nucleation and growth of many small crystals of the new phase. All these crystals, or grains, have the same structure but different orientations, so that, when they finally grow together, boundaries form between the grains. These boundaries play an important role in determining the properties of a piece of metal. At room temperature they strengthen the metal without reducing its ductility, but at high temperatures they often weaken the structure and lead to early failure. They can be the site of localized corrosion, which also leads to failure.

Mechanical Properties

When a metal rod is lightly loaded, the strain (measured by the change in length divided by the original length) is proportional to the stress (the load per unit of cross-sectional area). This means that, with each increase in load, there is a proportional increase in the rod's length, and, when the load is removed, the rod shrinks to its original size. The strain here is said to be elastic, and the ratio of stress to strain is called the elastic modulus. If the load is increased further, however, a point called the yield stress will be reached and exceeded. Strain will now increase faster than stress, and, when the sample is unloaded, a residual plastic strain (or elongation) will remain. The elastic strain at the yield stress is typically 0.1 to 1 percent, whereas, with the sample pulled to rupture, the plastic strain is typically 20 to 40 percent for an alloy (it may exceed 100 percent in some cases).

The most important mechanical properties of a metal are its yield stress, its ductility (measured by the elongation to fracture), and its toughness (measured by the energy absorbed in tearing the metal). The yield stress of a metal is determined by the resistance to slipping of one plane of atoms over another. Various barriers to slip can be produced by heat treatment and alloying; examples of such barriers are grain boundaries, fine precipitates, distortion introduced by cold working the metal, and alloying elements dissolved in the metal.

When a metal is made very strong through one or more of these methods, it may suddenly fracture under a load instead of yielding. This is particularly true when the metal contains notches or cracks that locally raise the stress and localize the yielding. The property of interest then becomes the fracture toughness, measured by the energy required to extend an existing crack in a piece of metal. In almost all cases, the fracture toughness of an alloy can be improved only by reducing its yield strength. The only exception to this is a smaller grain size, which increases both toughness and strength.

Magnetic Properties

When an electric current is passed through a coil of metal wire, a magnetic field is developed around the coil. When a piece of copper is placed inside the coil, this field increases by less than 1 percent, but, when a piece of iron, cobalt, or nickel is placed inside the coil, the external field can increase 10,000 times. This strong magnetic property is known as ferromagnetism, and the three metals listed above are the most prominent ferromagnetic metals. When the piece of ferromagnetic metal is removed from the coil, it retains some of this magnetism (that is, it is magnetized). If the metal is hard, as in a hardened piece of steel, the loss, or reversal, of magnetization will be slow, and the sample will be useful as a permanent magnet. If the metal is soft, it will quickly lose its magnetism; this will make it useful in electrical transformers, where rapid reversal of magnetization is essential.

In many types of solids, the atoms possess a permanent magnetic moment (they act like small bar magnets). In most solids, the direction of these moments is arranged at random. What is exceptional about ferromagnetic solids is that the interatomic forces cause the moments of neighbouring atoms spontaneously to align in the same direction. If the moments of all of the atoms in a single sample lined up in the same direction, the sample would be an exceptionally strong magnet with exceptionally high energy. That energy would be reduced if the sample broke up into domains, with all atomic moments in each domain being aligned but the direction of magnetization in adjacent domains being in opposite directions and thus tending to cancel one another. This is what happens when a ferromagnetic metal is magnetized: all domains do not take on the same orientation, but domains of one orientation grow at the expense of others. The alignment of atomic magnetic moments within a domain is weakened by thermally induced oscillations, and ferromagnetism is finally lost above the Curie point, which is 770 °C (1,420 °F) for iron and 358 °C (676 °F) for nickel.

ALLOYING METAL

Metals are sometimes used in their pure form; however more often than not, they are alloyed with other metals (or carbon) to give the metal more desirable properties (depending upon role). Transition metals form a wide range of alloys with each other. Their atoms are of a similar size and behaviour and so the lattice structure of the metal won't alter greatly as a result of substituting one atom for another. Even so, alloying metals modifies the properties of the metal and usually makes it harder and less malleable.

The Effect of Alloying on Metal Properties

Metallic bonds are strong but directed between particular atoms. When a force is applied to a metal crystal, the layers can slide past each other (slipping). After slipping has occurred, the atoms will stay in their new regular close-packed structure.

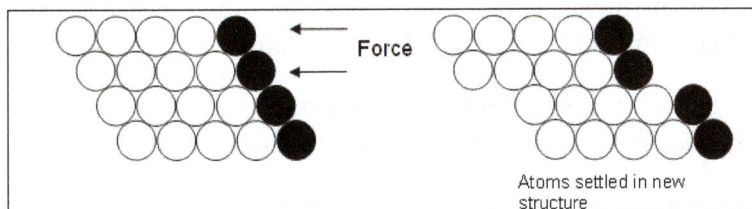

This is the reason why metals are malleable and ductile. When an alloy is added to the metal, the orderly arrangement of the lattice is disrupted; this prevents the metal layers from sliding past each other.

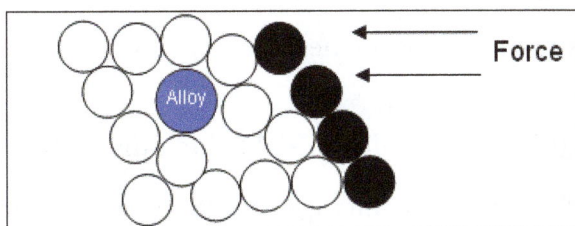

Smaller atoms can also be added into the metal lattice (often non-metals such as carbon or nitrogen); these fit into the holes between the atoms. This has the same effect of distorting the regular structure, thus making slip between the layers more difficult.

Steel Alloying Agents

Steel is essentially iron and carbon alloyed with certain additional elements. The process of alloying is used to change the chemical composition of steel and improve its properties over carbon steel or adjust them to meet the requirements of a particular application.

During the alloying process, metals are combined to create new structures that provide higher strength, less corrosion, or other properties. Stainless steel is an example of alloyed steel that includes the addition of chromium.

Benefits of Steel Alloying Agents

Different alloying elements each affect the properties of steel differently. Some of the properties that can be improved through alloying include:

- Stabilizing austenite: Elements such as nickel, manganese, cobalt, and copperincrease the temperatures range in which austenite exists.

- Stabilizing ferrite: Chromium, tungsten, molybdenum, vanadium, aluminum and silicon can help lower carbon's solubility in austenite. This results in an increase in the amount of carbides in the steel and decreases the temperature range in which austenite exists.

- Carbide forming: Many minor metals, including chromium, tungsten, molybdenum, titanium, niobium, tantalum and zirconium, form strong carbides that—in steel—increase hardness and strength. Such steels are often used to make high-speed steel and hot work tool steel.

- Graphitizing: Silicon, nickel, cobalt, and aluminum can decrease the stability of carbides in steel, promoting their breakdown and the formation of free graphite.

- Decrease of eutectoid concentration: Titanium, molybdenum, tungsten, silicon, chromium and nickel all lower the eutectoid concentration of carbon.

- Increase corrosion resistance: Aluminum, silicon and chromium form protective oxide layers on the surface of steel, thereby protecting the metal from further deterioration in certain environments.

Common Steel Alloying Agents

Below is a list of commonly used alloying elements and their impact on steel (standard content in parentheses):

- Aluminum (0.95-1.30%): A deoxidizer. Used to limit the growth of austenite grains.

- Boron (0.001-0.003%): A hardenability agent that improves deformability and machinability. Boron is added to fully killed steel and only needs to be added in very small quantities to have a hardening effect. Additions of boron are most effective in low carbon steels.

- Chromium (0.5-18%): A key component of stainless steels. At over 12 percent content, chromium significantly improves corrosion resistance. The metal also improves hardenability, strength, response to heat treatment and wear resistance.

- Cobalt: Improves strength at high temperatures and magnetic permeability.

- Copper (0.1-0.4%): Most often found as a residual agent in steels, copper is also added to produce precipitation hardening properties and increase corrosion resistance.

- Lead: Although virtually insoluble in liquid or solid steel, lead is sometimes added to carbon steels via mechanical dispersion during pouring in order to improve machinability.

- Manganese (0.25-13%): Increases strength at high temperatures by eliminating the formation of iron sulfides. Manganese also improves hardenability, ductility and wear resistance. Like nickel, manganese is an austenite forming element and can be used in the AISI 200 Series of Austenitic stainless steels as a substitute for nickel.

- Molybdenum (0.2-5.0%): Found in small quantities in stainless steels, molybdenum increases hardenability and strength, particularly at high temperatures. Often used in chromium-nickel austenitic steels, molybdenum protects against pitting corrosion caused by chlorides and sulfur chemicals.

- Nickel (2-20%): Another alloying element critical to stainless steels, nickel is added at over 8% content to high chromium stainless steel. Nickel increases strength, impact strength and toughness, while also improving resistance to oxidization and corrosion. It also increases toughness at low temperatures when added in small amounts.

- Niobium: Has the benefit of stabilizing carbon by forming hard carbides and is often found in high-temperature steels. In small amounts, niobium can significantly increase the yield strength and, to a lesser degree, the tensile strength of steels as well as have a moderate precipitation strengthening the effect.

- Nitrogen: Increases the austenitic stability of stainless steels and improves yield strength in such steels.

- Phosphorus: Phosphorus is often added with sulfur to improve machinability in low alloy steels. It also adds strength and increases corrosion resistance.

- Selenium: Increases machinability.

- Silicon (0.2-2.0%): This metalloid improves strength, elasticity, acid resistance and results in larger grain sizes, thereby, leading to greater magnetic permeability. Because silicon is used in a deoxidizing agent in the production of steel, it is almost always found in some percentage in all grades of steel.

- Sulfur (0.08-0.15%): Added in small amounts, sulfur improves machinability without resulting in hot shortness. With the addition of manganese hot shortness is further reduced due to the fact that manganese sulfide has a higher melting point than iron sulfide.

- Titanium: Improves both strength and corrosion resistance while limiting austenite grain size. At 0.25-0.60 percent titanium content, carbon combines with the titanium, allowing chromium to remain at grain boundaries and resist oxidization.

- Tungsten: Produces stable carbides and refines grain size so as to increase hardness, particularly at high temperatures.

- Vanadium (0.15%): Like titanium and niobium, vanadium can produce stable carbides that increase strength at high temperatures. By promoting a fine grain structure, ductility can be retained.

- Zirconium (0.1%): Increases strength and limits grains sizes. Strength can be notably increased at very low temperatures (below freezing). Steel's that include zirconium up to about 0.1% content will have smaller grains sizes and resist fracture.

ELECTRICAL CONDUCTIVITY OF METALS

Electrical conductivity in metals is a result of the movement of electrically charged particles. The atoms of metal elements are characterized by the presence of valence electrons, which are electrons in the outer shell of an atom that are free to move about. It is these 'free electrons' that allow metals to conduct an electric current.

Because valence electrons are free to move they can travel through the lattice that forms the physical structure of a metal. Under an electric field, free electrons move through the metal much like billiard balls knocking against each other, passing an electric charge as they move.

Transfer of Energy

The transfer of energy is strongest when there is little resistance. On a billiard table, this occurs when a ball strikes against another single ball, passing most of its energy onto the next ball. If a single ball strikes multiple other balls, each of those will carry only a fraction of the energy.

By the same token, the most effective conductors of electricity are metals that have a single valence electron that is free to move and causes a strong repelling reaction in other electrons. This is the case in the most conductive metals, such as silver, gold, and copper, who each have a single valence electron that moves with little resistance and causes a strong repelling reaction.

Semiconductor metals (or metalloids) have a higher number of valence electrons (usually four or more) so, although they can conduct electricity, they are inefficient at the task. However, when heated or doped with other elements semiconductors like silicon and germanium can become extremely efficient conductors of electricity.

Metal Conductivity

Conduction in metals must follow Ohm's law, which states that the current is directly proportional to the electric field applied to the metal. The law, named after German physicist Georg Ohm, appeared in 1827 in a published paper laying out how current and voltage are measured via electrical circuits. The key variable in applying Ohm's law is a metal's resistivity.

Resistivity is the opposite of electrical conductivity, evaluating how strongly a metal opposes the flow of electric current. This is commonly measured across the opposite faces of a one-meter cube of material and described as an ohm meter ($\Omega \cdot$m). Resistivity is often represented by the Greek letter rho (ρ).

Electrical conductivity, on the other hand, is commonly measured by siemens per meter (S·m^{-1}) and represented by the letter sigma (σ). One siemens is equal to the reciprocal of one ohm.

Table: Conductivity and Resistivity of Metals.

Material	Resistivity $\rho(\Omega \cdot$m) at 20 °C	Conductivity σ(S/m) at 20 °C
Silver	1.59×10^{-8}	6.30×10^7
Copper	1.68×10^{-8}	5.98×10^7
Annealed Copper	1.72×10^{-8}	5.80×10^7
Gold	2.44×10^{-8}	4.52×10^7
Aluminum	2.82×10^{-8}	3.5×10^7
Calcium	3.36×10^{-8}	2.82×10^7
Beryllium	4.00×10^{-8}	2.500×10^7
Rhodium	4.49×10^{-8}	2.23×10^7
Magnesium	4.66×10^{-8}	2.15×10^7

Molybdenum	5.225×10^{-8}	1.914×10^{7}
Iridium	5.289×10^{-8}	1.891×10^{7}
Tungsten	5.49×10^{-8}	1.82×10^{7}
Zinc	5.945×10^{-8}	1.682×10^{7}
Cobalt	6.25×10^{-8}	1.60×10^{7}
Cadmium	6.84×10^{-8}	1.46^{7}
Nickel (electrolytic)	6.84×10^{-8}	1.46×10^{7}
Ruthenium	7.595×10^{-8}	1.31×10^{7}
Lithium	8.54×10^{-8}	1.17×10^{7}
Iron	9.58×10^{-8}	1.04×10^{7}
Platinum	1.06×10^{-7}	9.44×10^{6}
Palladium	1.08×10^{-7}	9.28×10^{6}
Tin	1.15×10^{-7}	8.7×10^{6}
Selenium	1.197×10^{-7}	8.35×10^{6}
Tantalum	1.24×10^{-7}	8.06×10^{6}
Niobium	1.31×10^{-7}	7.66×10^{6}
Steel (Cast)	1.61×10^{-7}	6.21×10^{6}
Chromium	1.96×10^{-7}	5.10×10^{6}
Lead	2.05×10^{-7}	4.87×10^{6}
Vanadium	2.61×10^{-7}	3.83×10^{6}
Uranium	2.87×10^{-7}	3.48×10^{6}
Antimony*	3.92×10^{-7}	2.55×10^{6}
Zirconium	4.105×10^{-7}	2.44×10^{6}
Titanium	5.56×10^{-7}	1.798×10^{6}
Mercury	9.58×10^{-7}	1.044×10^{6}
Germanium*	4.6×10^{-1}	2.17
Silicon*	6.40×10^{2}	1.56×10^{-3}

The resistivity of semiconductors (metalloids) is heavily dependent on the presence of impurities in the material.

THERMAL CONDUCTIVITY OF METALS

Thermal Conductivity is a term analogous to electrical conductivity with a difference that it concerns with the flow of heat unlike current in the case of the latter. It points to the ability of a material to transport heat from one point to another without movement of the material as a whole, the more is the thermal conductivity the better it conducts the heat.

Let us consider a block of material with one end at temperature T_1 and other at T_2. For $T_1 > T_2$, heat flows from T_1 end to T_2 end, and the heat flux(J) flowing across a unit area per unit time is given as:

$$\frac{\Delta Q}{A \Delta t} = J = -K \frac{dt}{dx}$$

Where, K is the thermal conductivity in Joule/meter-sec-K or Watts/meter-K.

Generally the heat transfer in solid has two components:

- Lattice conduction.

- Electronic conduction.

Both types of heat conduction occur in solids but one is dominant over the other depending upon the type of material. In case of insulating materials, lattice conduction contributes to heat conduction. This is mainly due to the fact that in insulators the electrons are tightly held by their parent atoms and free electrons do not exist. Hence the heat is transferred from one end to another by vibration of atoms held in the lattice structure. Obviously insulators are bad heat conductors since they do not posses enough heat transfer capability due to lack of free electrons.

However in case of metals we have large number of free electrons, and hence heat conduction is primarily due to electronic conduction. The free electrons of the metals can freely move throughout the solid and transfer the thermal energy at a rate very high as compared to insulators. It is due to this that the metals posses high thermal conductivity. It is also observed that among the metals the best electrical conductors also exhibits best thermal conductivity. Since both electrical as well as thermal conductivity is dependent on the free electrons, factors such as alloying effect both the properties. Thermal conductivity of metals vary from 15 – 450 W/mK at 300K.

Wiedemann Franz Law

Wiedemann Franz law basically relates the two conductivities of metals, i.e. thermal and electrical conductivity with temperature. It states that the ratio of thermal conductivity K and electrical conductivity σ is proportional to the temperature of the specimen. G. Wiedemann and R. Franz in 1853 established on the basis of experimental data that the ratio $\dfrac{K}{\sigma}$ is constant at constant temperature.

In 1882 the Danish physicist L. Lorenz demonstrated that the relation $\dfrac{K}{\sigma}$ changes in direct proportion to the absolute temperature T.

$$\frac{K}{\sigma} = LT$$

Where, T = temperature

$$L \stackrel{=}{} 2.54 \times 10^{-8} W\Omega / K^2, \textit{Lorentz number} \left(a \text{ constant} \right)$$

This law basically states that with increase in temperature the thermal conductivity of metals increases while the electrical conductivity decreases. We know that the two properties of metals are dependent on the free electrons. An increase in temperature increases the average velocity of the free electrons leading to increase in heat energy transfer. On the other hand increase in the velocity of electrons also increases the number of collisions of the free electrons with lattice ions, and hence contributes to increase in electrical resistivity or reduction in electrical conductivity.

However this law has certain limitations. The proportionality does not holds true for all ranges of temperature. It is only found valid for very high temperatures and very low temperatures. Also certain metals such as beryllium, pure silver etc. do not follow this law.

References

- Physical-metallurgy, metallurgy, science: britannica.com, Retrieved 12 April, 2019

- Alloy: 4college.co.uk, Retrieved 13 May, 2019

- Common-steel-alloying-agents-properties-and-effects: thebalance.com, Retrieved 14 June, 2019

- Electrical-conductivity-in-metals: thebalance.com, Retrieved 15 July, 2019

- Thermal-conductivity-of-metals: electrical4u.com, Retrieved 16 August, 2019

Powder Metallurgy

Powder metallurgy is the science of creating various materials and compounds from metal powder. It includes the processes such as powder production, mixing powders for PM processing, sintering, etc. The chapter closely examines these processes and properties of powder metallurgy to provide an extensive understanding of the subject.

Powder metallurgy is a kind of manufacturing process in which we make final products with the help of powdered metal. In this process we heat powdered metal to a temperature below its melting point and then compact that metal to a desired shape.

Since the powdered metal gets compacted it its final shape so we do not need any further machining to obtain final product.

Sintering means heating of green compact in an oven.

Iron powder.

Advantages of Powder Metallurgy

- Products made by P/M generally do not require further finishing.

- There is no wastage of raw material.

- Reasonably complex shapes can be made.

- Different combinations of materials can be used in P/M products, which are otherwise impossible to make. For example, mixing ceramics with metals.

- Automation of P/M process is easy as compared to other manufacturing processes.
- It provides properties like porosity and self-lubrication to the manufactured parts.

Limitations of Powder Metallurgy

- Tooling cost is generally and can only be justified in mass production.
- Raw material cost is very high.
- Mechanical properties of the parts are of low quality as compared to cast or machined parts.
- In some cases, density of different parts of final product can very due to uneven compression.
- Size of product that can be manufactured is generally limited to 2-20 kgs.

Applications of Powder Metallurgy

- P/M parts are generally used as filters due to porous nature.
- Making cutting tools and dies.
- Used in making machinery parts.
- Due to self-lubricating property P/M components are widely used in making bearings and bushes.
- P/M process is also used in making magnets.

PROPERTIES OF POWDER METALLURGY MATERIALS

The mechanical properties available from the materials, commonly used for structural or engineering component applications, can be summarized as follows:

Ferrous Powder Metallurgy Materials

Ferrous Powder Metallurgy materials, processed by the standard die press and sinter route, can deliver UTS levels up to around 900 N/mm^2 in the as-sintered condition or up to around 1200 N/mm^2 after heat treatment or sinter hardening.

These pressed and sintered materials can also deliver tensile yield stress levels up to around 480 N/mm2 as-sintered or around 1200 N/mm^2 after heat treatment or sinter hardening. Compressive yield stresses are slightly higher at up to around 510 N/mm^2 as sintered or up to around 1250 N/mm^2 heat treated.

These very significant levels of strength are, however, accompanied by quite low levels of tensile ductility (Elongation levels below 2% being quite typical). For this reason, PM products at conventional press/sinter density levels (up to 7.1-7.2 g/cm³ maximum) would not be used in applications likely to experience gross plasticity in service.

Powder Forged Steels

Powder forged steels can deliver high strength levels (UTS up to around 950 N/mm² as forged and 2050 N/mm² heat treated; tensile yield stress up to around 650 N/mm² as forged and 1760 N/mm² heat treated) with higher levels of ductility (5-18% Elongation).

Stainless Steels

300 series PM stainless steels, in the Press/Sinter condition, can deliver UTS levels up to around 480 N/mm², tensile yield stress up to around 310 N/mm²and compressive yield strength up to around 320 N/mm², but with much higher ductility levels than their low alloy steel counterparts. (>10% Elongation).

400 series PM stainless steels can deliver similar properties to the 300 series materials in the as-sintered condition. Heat treatment of martensitic grades can increase strength levels to up to around 720 N/mm² UTS and tensile yield stress and 640 N/mm² compressive yield stress, but at the expense of a much reduced ductility (<1% Elongation).

Copper Alloys

Press and sintered Cu alloys can deliver relatively modest strength levels (up to around 240 N/mm² UTS, 140 N/mm² tensile yield stress and 170 N/mm² compressive yield stress) but with much higher ductility than their ferrous counterparts (10-20% Elongation).

Aluminium Alloys

Pressed and sintered Al alloys can deliver UTS of up to around 200 N/mm2 as sintered or up to around 320 N/mm² after heat treatment and tensile yield stress of up to around 170 N/mm² as sintered or up to around 320 N/mm² after heat treatment, but with quite low ductility levels (0.5-2% Elongation).

Fatigue Strength

Both Press/Sinter Powder Metallurgy steels and Powder Forged steels are capable of providing significant levels of fatigue strength:

- In the as-sintered condition, Press/Sinter PM steels can deliver fatigue endurance limits of up to around 320 N/mm² in the rotating bend loading mode and up to around 270 N/mm² in the axial loading mode (R = -1, Kt = 1).

- Heat treatment can raise these values to up to around 540 N/mm² and 460 N/mm² respectively.

- Powder forged steels can deliver fatigue endurance limits of up to around 420 N/mm² in the rotating bend loading mode and up to around 360 N/mm² in the axial loading mode (R = -1, Kt = 1).

- Heat treatment can raise these values to up to around 635 N/mm2 and 560 N/mm² respectively.

Global Powder Metallurgy Property Database

Design Engineers and other material specifiers in organisations that are customers or potential customers for Powder Metallurgy components, ought to be aware of the Global Powder Metallurgy Property Database.

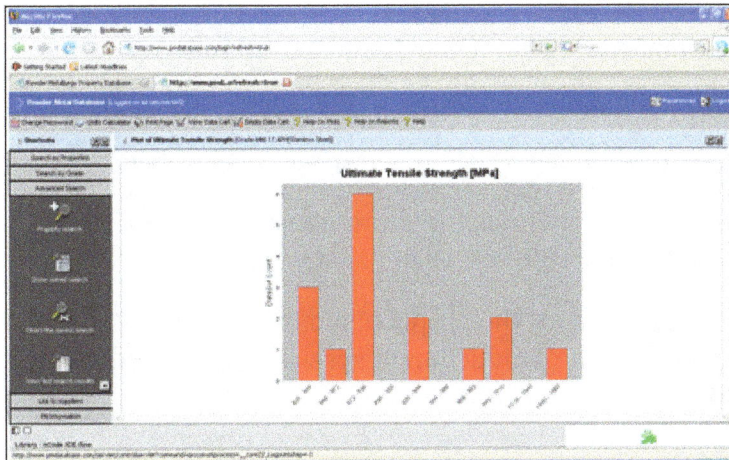

Typical results obtained from the GPMPD, this example showing tensile strength.

This database is a very significant resource for anyone assessing the capabilities of Powder Metallurgy for target applications and holds verified and accredited mechanical, fatigue and physical properties across a number of material categories:

- Powder Metallurgy low alloy steels.

- Stainless steels.

- Non-ferrous structural part materials.

- Powder Forging steels.

- Powder Metallurgy self-lubricating bearing alloys.

- Metal Injection Moulding alloys – ferrous and non-ferrous.

Full fatigue SN-curve information is also downloadable, in pdf format, for a range of Fe-Cu-C PM structural part grades, tested under a range of fatigue conditions.

POWDER PRODUCTION

In tonnage terms, the production of iron powders for PM structural part production dwarfs the production of all of the non-ferrous metal powders combined. Virtually all iron powders are produced by one of two processes.

The Sponge Iron Process

The longest established of these processes is the sponge iron process, the leading example of a family of processes involving solid state reduction of an oxide. In the process, selected magnetite (Fe_3O_4) ore is mixed with coke and lime and placed in a silicon carbide retort.

The filled retort is then passed through a long kiln, where the reduction process leaves an iron "cake" and a slag. In subsequent steps, the retort is emptied, the reduced iron sponge is separated from the slag and is crushed and annealed.

The resultant powder is highly irregular in particle shape, therefore ensuring good "green strength" so that die-pressed compacts can be readily handled prior to sintering, and each particle contains internal pores (hence the term "sponge") so that the good green strength is available at low compacted density levels.

Sponge iron provides the base feedstock for all iron-based, self-lubricating bearings and still accounts for around 30% of iron powder usage in PM structural parts.

Solid state reduction is also used for the production of refractory metal powders, using hydrogen as the reducing agent, and for the production of specialist iron powders by the reduction of mill scale (again using hydrogen).

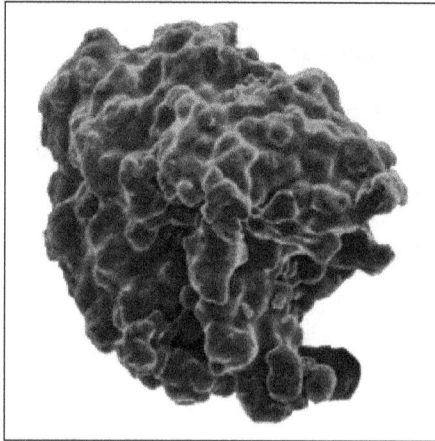

Typical morphology of an iron powder particle (ASC100.29) produced by water atomisation.

Water Atomisation

Driven by the trend towards higher density levels in PM structural parts as a means of increasing performance levels, sponge iron powders have been increasingly supplanted by powders made by water atomisation.

Atomisation involves the disintegration of a thin stream of molten metal through the impingement of high energy jets of a fluid (liquid or gas). Water is the most commonly used liquid in atomisation.

Water atomised iron powders also have irregular particle shape and therefore good green strength. Unlike sponge iron, the individual powder particles do not contain internal porosity and, because of extensive development of the annealing process, have superior compressibility. Water atomised powders are therefore the material of choice where high green density is sought in PM structural parts.

Non-ferrous Metal Powder Production

Inert Gas Atomisation

Non-ferrous metal powders are produced by a variety of means. The most significant of these is another atomising process, this time using an inert gas as the atomising fluid. In inert gas atomisation, the particle shape produced is dependent on the time available for surface tension to take effect on the molten droplets prior to solidification and, if a low heat capacity gas is used (nitrogen and argon are most common), this time is extended and spherical powder shapes result.

Spherical powders are particularly useful in hot isostatic pressing, where green strength is not an issue but initial packing density of the powder in the container is significant.

Close-coupled Atomisation

The atomising nozzle design can provide either free-fall or close-coupled atomisation. In close-coupled (or confined) atomisation, the design of pouring nozzle and atomising head is adjusted so that impingement of the gas jets and molten stream occurs immediately below the exit of the nozzle with little or no free-fall height. This variant of atomisation technology has proved particularly useful for the production of fine powders for a range of applications, including Metal Injection Moulding.

Plasma Atomisation

The plasma atomisation process uses argon plasma torches at > 10,000 °C to melt and atomise titanium and other metals into fine droplets. The process has the distinction of producing highly flowable and very pure spherical metallic powders using wire as its feedstock. This method ensures a high level of traceability allowing for applications in the biomedical and aerospace sectors.

Centrifugal Atomisation

A further branch of the "atomisation family" comprises a number of centrifugal atomisation processes. There are essentially two types of such processes; in the first type, a cup of molten metal is rotated at high speed or a molten stream of metal is allowed to fall onto a rotating disc or cone; in the second type, the Rotating Electrode Process (REP), a bar of metal is rotated and the free end is progressively melted by an arc from a tungsten electrode. If a plasma arc is involved, the process is known as PREP (Plasma Rotating Electrode Process) and this is a leading candidate for titanium powder production.

There are a few other powder production technologies that have areas of application.

Electrolysis

Electrolysis is a means of producing metal powders and has been most commonly used for the manufacture of copper powders for specialist applications. Electrolytic powders are produced by following the principles used in electroplating, with the conditions changed to produce a loose powdery deposit rather than a smooth adherently solid layer. The formation of powder deposits that adhere loosely to the cathode is favoured by low metal ion concentration in the electrolyte, high acid concentration and high cathode current density. The starting material is a pure metal anode.

Mechanical Comminution

Brittle materials can be pulverised in ball mills, hammer mills or attritor mills to form powders. Intermetallics and ferro-alloys are commonly processed this way. As variants on this approach, Hydride-Dehydride (HDH) titanium alloy powders can be produced by reacting the alloy in solid form with hydrogen to form a brittle hydride, which can then be pulverised and dehydrided, and Hydrogen Decrepitation of Nd-Fe-B magnetic alloys, which can cause spontaneous decrepitation of the solid alloy.

Carbonyl and Chemical Conversion

Finally, there is a range of chemical conversion processes, with the leading example being the carbonyl process for the production of fine nickel or iron powders. In this process, the crude metal is reacted with CO under pressure to form the carbonyl, which is gaseous at reaction temperature, but decomposes to deposit the metal on raising temperature and lowering pressure.

Other chemical conversion processes include:

- The manufacture of Platinum powders from sponge created by thermally decomposing platinum ammonium chloride.

- The Sherritt-Gordon process for the manufacture of nickel powders by hydrogen reduction of a solution of a nickel salt under pressure.

- Chemical precipitation of metals from solution of a soluble salt e.g. silver can be precipitated by adding a reducing agent to a silver nitrate solution.

MIXING POWDERS FOR PM PROCESSING

The mixing, or blending, of powder feedstocks for die pressing of Powder Metallurgy parts is carried out for two reasons:

To Introduce Alloying Element Additions in a Homogeneous Form

Die pressing feedstocks generally consist of elemental mixes in order to maintain as high a level of compressibility as possible. Using this approach means that the compressibility is controlled by that of the soft, annealed base powder (most commonly iron). Use of a fully pre-alloyed powder would mean that the initial particle hardness and work hardening rate would both be increased by the alloying additions and compressibility therefore reduced.

To Incorporate a Pressing Lubricant

Popular lubricants are stearic acid, stearin, metallic stearates or other organic compound of a waxy nature. The purposes of adding the lubricant are to reduce friction (and therefore even out density variations) during compaction, to reduce ejection forces and to minimise the tendency for ejection cracking in the compact.

A homogeneous mix is generally produced from the initial constituents by a tumbling action in an appropriate mixing vessel. Mixing vessels are often of a double-cone geometry, but other vessel shapes are also utilised (V, W or Y-shaped sections).

In the special case of cemented carbide materials, mixing is carried out in a ball mill, in order to coat the individual carbide particles with the binder metal (e.g. cobalt). As the very fine powder particles involved have poor flow characteristics, the mixture is subsequently granulated to form agglomerates.

POWDER METALLURGY COMPACT

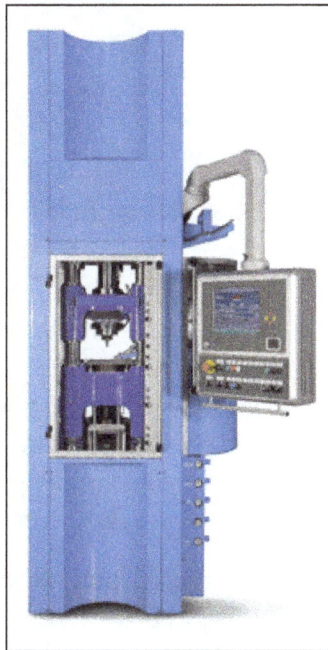

Direct electrical drive press for hardmetal compaction.

The dominant technology for the forming of products from powder materials, in terms of both tonnage quantities and numbers of parts produced, is Die Pressing. This forming technology involves a production cycle comprising:

- Filling a die cavity with a known volume of the powder feedstock, delivered from a fill shoe.

- Compaction of the powder within the die with punches to form the compact. Generally, compaction pressure is applied through punches from both ends of the toolset in order to reduce the level of density gradient within the compact.

- Ejection of the compact from the die, using the lower punches.

- Removal of the compact from the upper face of the die using the fill shoe in the fill stage of the next cycle.

This cycle offers a readily automated and high production rate process. There are, however, some limitations to the products delivered by this route:

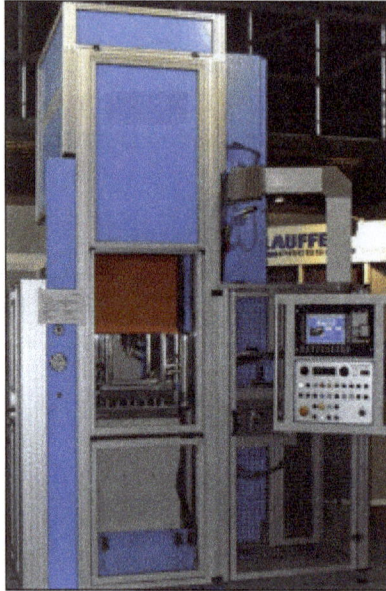

Structural part press of 630 kN capacity for eight tool levels; all hydraulic pistons are located in upper ram and lower crosshead of the press frame.

Geometrical Complexity

The geometrical complexity that can be delivered might best be described as "two and half dimensional". There is unlimited complexity in the radial directions (i.e. in the plan view of the part); if the shape can be cut into the die, then it can be formed in the part. In the third dimension, the axial or through-thickness direction of the part, there are, however, significant limitations.

Changes in section thickness can be created by the use of multiple top and bottom punches and holes in this direction can be created through the incorporation of core rods and mandrels in the toolset. However, re-entrant features cannot be formed as they would impede ejection of the part from the die.

Aspect Ratio

The aspect (length to diameter) ratio of the part is limited (to around 3:1) if acceptable control over density variations is sought.

Size and Weight

The size and weight of the part is limited by the maximum tonnage capacity of available forming presses (around 1000 tonnes capacity). A 2 kg. ferrous PM part would be regarded as being a large one.

Strength

The strength level of conventionally die pressed parts is limited to some degree by the influence of the remaining porosity in the product. A range of process developments have been introduced, many of them evolutions of the standard press/sinter process, which have tackled this particular issue.

A number of alternative forming processes have been developed, which have sought to attack one or more of these limitations.

Isostatic Pressing

Isostatic pressing can tackle all four limitations in that very large components can be formed, the only limit on achievable aspect ratio arises from the dimensions of the vessel containing the pressing fluid, true three-dimensional geometrical complexity can be achieved and full-density compaction can be delivered. All this is, however, at the expense of significant increases in forming cycle time and some limitations in dimensional tolerance control, compared with die pressing.

In isostatic pressing, the powder is compacted with a hydrostatic pressure in all directions. The process can be carried out cold or hot.

Cold Isostatic Pressing

In cold isostatic pressing, the powder is contained in a flexible mould, commonly of polyurethane, which is immersed in liquid, usually water, in a pressure vessel, which is pumped to high pressure.

Hot Isostatic Pressing

In hot isostatic pressing, the pressuring medium is a gas, normally argon. The powder is contained in a metallic can, which is subjected to the hydrostatic pressure in the pressure vessel. Full density is achievable by HIP and the process is used for superalloys, high speed steels, titanium etc. where integrity of the materials is a prime consideration.

A significant contributor to HIP costs arises from the canning process. So, there is significant interest in "canless HIP" processing. If the powder can be consolidated to a density above about 92% by a preliminary forming process (e.g. die pressing or CIP), surface connected porosity can be eliminated, gas penetration into the part can be avoided during the subsequent HIP process and full densification can be achieved.

A variant of this approach is Sinter-HIP, in which the required 92%+ density is achieved by sintering and then HIP consolidation is applied in the same vessel.

Split Die Compaction

An early process development aimed at enhancing geometrical complexity of PM parts and introducing the capability for forming re-entrant features was an evolutionary extension of die pressing, known as split die compaction. In this process, the toolset was engineered so that, after the part was formed, the die could be parted in the horizontal plane, allowing the compact with the re-entrant features to be ejected between the two halves of the die.

Metal Injection Moulding

However, the process that has had the major impact in extending shape capability of parts made from powders has been Metal Injection Moulding (MIM). In MIM, the powder is mixed with an organic binder to create a feedstock, which can be injected into a mould in a manner identical to the process used for injection moulded plastic parts. After release of the green body from the mould, it is debound and then sintered.

Close dimensional tolerances can be held in this process and, because of the fine powders used and consequent high sintering activity, density levels close to full density and therefore high strength levels can be achieved. Also, extremely complex 3-dimensional geometries can be formed. However, principally because of the difficulties involved with the debinding stage of the process, MIM parts are generally small and light with thin-walled sections. As a comparison with die-pressed parts, a 100 g. MIM part would be considered large.

SINTERING

Sintering is a heat treatment applied to a powder compact in order to impart strength and integrity. The temperature used for sintering is below the melting point of the major constituent of the Powder Metallurgy material.

After compaction, neighbouring powder particles are held together by cold welds, which give the compact sufficient "green strength" to be handled. At sintering temperature, diffusion processes cause necks to form and grow at these contact points.

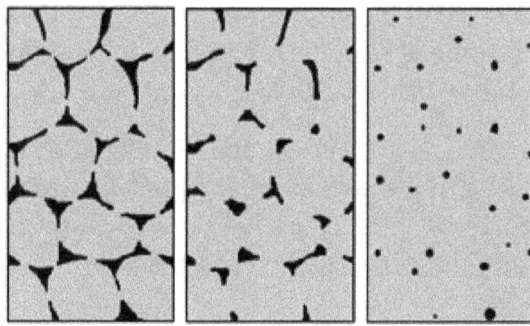

The three stages of solid state sintering: left: initial stage, centre: intermediate stage, right: final stage.

There are two necessary precursors before this "solid state sintering" mechanism can take place:

- Removal of the pressing lubricant by evaporation and burning of the vapours.

- Reduction of the surface oxides from the powder particles in the compact.

These steps and the sintering process itself are generally achieved in a single, continuous furnace by judicious choice and zoning of the furnace atmosphere and by using an appropriate temperature profile throughout the furnace.

Sinter Hardening

Sintering furnaces are available that can apply accelerated cooling rates in the cooling zone and material grades have been developed that can transform to martensitic microstructures at these cooling rates. This process, together with a subsequent tempering treatment, is known as sintering hardening, a process that has emerged, in recent years, has a leading means of enhancing sintered strength.

Liquid Phase Sintering

Transient Liquid Phase Sintering

In a compact that contains only iron powder particles, the solid state sintering process would generate some shrinkage of the compact as the sintering necks grow. However, a common practice with ferrous PM materials is to make an addition of fine copper powder to create a transient liquid phase during sintering.

At sintering temperature, the copper melts and then diffuses into the iron powder particles creating swelling. By careful selection of copper content, it is possible to balance this swelling against the natural shrinkage of the iron powder skeleton and provide a material that does not change in dimensions at all during sintering. The copper addition also provides a useful solid solution strengthening effect.

Permanent Liquid Phase Sintering

For certain materials, such as cemented carbides or hardmetals, a sintering mechanism involving the generation of a permanent liquid phase is applied. This type of liquid phase sintering involves the use of an additive to the powder, which will melt before the matrix phase and which will often create a so-called binder phase. The process has three stages:

- Rearrangement:

  As the liquid melts, capillary action will pull the liquid into pores and also cause grains to rearrange into a more favourable packing arrangement.

- Solution-precipitation:

  In areas where capillary pressures are high, atoms will preferentially go into solution and then precipitate in areas of lower chemical potential where particles are not close or in contact. This is called contact flattening and densifies the system in a way similar to grain boundary diffusion in solid state sintering. Ostwald ripening will also occur where smaller particles will go into solution preferentially and precipitate on larger particles leading to densification.

- Final densification:

  Densification of the solid skeletal network, liquid movement from efficiently packed regions into pores. For permanent liquid phase sintering to be practical, the major phase should be at least slightly soluble in the liquid phase and the "binder" additive should melt before any major sintering of the solid particulate network occurs, otherwise rearrangement of grains will not occur.

References

- Powder-metallurgy: clubtechnical.com, Retrieved 17 January, 2019

- Properties-of-powder-metallurgy-materials: pm-review.com/introduction-to-powder-metallurgy, Retrieved 18 February, 2019

- Powder-production-technologies, introduction-to-powder-metallurgy: pm-review.com, Retrieved 19 March, 2019

- Mixing-powders-for-pm-processing, introduction-to-powder-metallurgy: pm-review.com, Retrieved 20 April, 2019

- Forming-a-powder-metallurgy-compact, introduction-to-powder-metallurgy: pm-review.com, Retrieved 21 May, 2019

- Sintering-in-the-powder-metallurgy-process, introduction-to-powder-metallurgy: pm-review.com, Retrieved 22 June, 2019

Metalworking

The process of working with metals to create individual parts or large structures is known as metalworking. Some of its components consist metal forming, molding, extrusion, metal casting, metal cutting, metal spinning, etc. This chapter delves into study of these components to provide in-depth knowledge for the subject.

Metalworking is the process of working with metals to create individual parts, assemblies, or large-scale structures.

The term 'metalworking' covers a huge variety of work across multiple industries/sectors from companies making large ships and bridges through to precision metal manufacturers.

The Most Common Metalworking Processes

Because metalworking is used so diversely it includes a wide range of skills, processes and tools. Most commonly metalworking is divided into the following categories:

- Cutting: Cutting is a collection of processes where material is brought to a specified shape by removing excess using various kinds of tooling.

- Forming: Forming modifies metal by deforming it i.e. forming does not remove any metal. Forming is done with a system of mechanical forces and, especially for bulk metal forming, with heat.

- Joining: Joining brings two or more pieces of metal together by way of one or more different processes which include welding, brazing and soldering.

Commonly used Metalworking Tools

There are many, many different commonly used metalworking tools ranging from the simple (but effective) hammer and anvil through to complex tools including Fibre Lasers, punch presses, Salvagnini P4 panel benders, Amada Press Breaks and, MIG and TIG welding equipment – kit that we have here at JC Metalworks.

Of course working with metalworking tools are the people that operate them and their skills are as important as the tools and equipment they operate.

METAL FORMING

Metal forming processes, also known as mechanical working processes, are primary shaping processes in which a mass of metal or alloy is subjected to mechanical forces. Under the action of such forces, the shape and size of metal piece undergo a change. By mechanical working processes, the given shape and size of a machine part can be achieved with great economy in material and time. Metal forming is possible in case of such metals or alloys which are sufficiently malleable and ductile. Mechanical working requires that the material may undergo "plastic deformation" during its processing. Frequently, work piece material is not sufficiently malleable or ductile at ordinary room temperature, but may become so when heated. Thus we have both hot and cold metal forming operations.

When a single crystal is subjected to an external force, it first undergoes elastic deformation; that is, it returns to its original shape when the force is removed. For example, the behavior is a helical spring that stretches when loaded and returns to its original shape when the load is removed. If the force on the crystal structure is increased sufficiently, the crystal undergoes plastic deformation or permanent deformation; that is, it does not return to its original shape when the force is removed.

There are two basic mechanisms by which plastic deformation takes place in crystal structures. One is the slipping of one plane of atoms over an adjacent plane (called the slip plane) under a shear stress. The behavior is much like the sliding of playing cards against each other. Shear stress is defined as: The ratio of the applied shearing force to the cross-sectional area being sheared, just as it takes a certain magnitude of force to slide playing cards against each. In other word we can say that a single crystal requires a certain amount of shear stress (called critical shear stress) to undergo permanent deformation. Thus, there must be a shear stress of sufficient magnitude within a crystal for plastic deformation to occur; otherwise the deformation remains elastic.

The second and less common mechanism of plastic deformation in crystals is twinning, in which a portion of the crystal forms a mirror image of itself across the plane of twinning. Twins form abruptly and are the cause of the creaking sound ("tin cry") that occurs when a tin or zinc rod is bent at room temperature. Twinning usually occurs in hcp metals.

Yield Criteria

The yield criteria limit the elastic region. It is a mathematical expression to define the combination of component of stress such that when it reaches material no more behaves elastically. Yield criterion gives the onset plastic deformation. In other word if a state of stress satisfies yield criterion, we can say that plastification may start. It is assumed that initial yielding depends upon only on state of stress and not on how the stress is reached. We can assume that there exist a function $f\left(\sigma_{ij}\right)$ called yield function such that:

Material is elastic if: $f\left(\sigma_{ij}\right) < 0$,

or if $f\left(\sigma_{ij}\right) = 0$ and $f\left(\sigma_{ij}\right) < 0$,

where $f\left(\sigma_{ij}\right) = 0$ defines the yield surface in stress space and $f\left(\sigma_{ij}\right) = 0$ indicates unloading. The latter combination tells us the onset plastification has taken place, but unloading is going to take place elastically. As the yield criterion does not depends upon the path of loading, it does not tell anything about deformation. If the state of stress is already satisfied $f(\sigma_{ij})=0$, it tells us only the plastifiaction has just started or taken place. But it does not tell whether plastic deformation has taken place or not. The yield function gives us the information regarding loading.

Material behavior is plastic if:

$$f(\sigma_{ij})=0 \text{ or } f(\sigma_{ij})\geq 0$$

Commonly used Yield Criteria

The yield criteria of materials limit the elastic domain during loading where as the failure criteria gives the maximum stress that can be applied. We use the yield criteria for metals alloys and failure criteria for geo material like soil and concrete. Some of the commonly used yield criteria are:

- Von Mises yield criteria.

- Tresca yield criteria.

Von Mises Yield Criteria

Von Mises suggested that yielding will occur when second invariants of deviatoric stress tensor, J2 reaches a critical value. He does not take J3 into account in the yield criteria. We can write the at onset of yielding.

$$2J_2 = S_{ij}S_{ij} = S_1^2 + S_2^2 + S_3^2 = 2K^2$$

Where S_1, S_2, S_3 are principal deviator stress. We can also write von mises criteria in terms of principal stresses as $(\sigma_1-\sigma_2)^2+(\sigma_2-\sigma_3)^2+(\sigma_3-\sigma_1)^2 = 6k^2$.

In terms of components of stress tensor,von Mises yield criteria can be written as:

$$(\sigma_x-\sigma_y)^2+(\sigma_y-\sigma_z)^2+(\sigma_z-\sigma_x)^2+6(\tau_{yz}^2+\tau_{zx}^2+\tau_{xy}^2)=6k^2$$

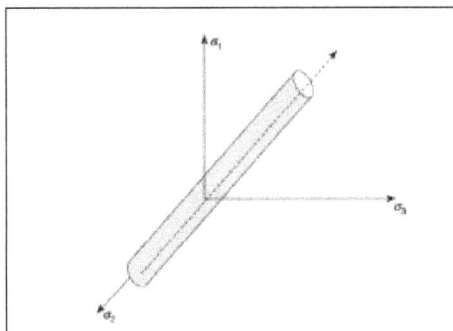

Let effective stress σ_{eff} corresponding to stress tensor σ as:

$$\sigma_{eff}=\sqrt{\frac{3}{2}S_{ij}S_{ij}}=\sqrt{\frac{3}{2}S:S}$$

Where S_{ij} is the components of deviatoric stress tensor S.von Mises criteria can be written as $\sigma_{eff} - \sigma_y = 0$ where σ_y is the yield stress of the material in uniaxial tension or compression.

Tresca Yield Criteria

According to the tresca yield criteria, yielding of material begin to occur when maximum shearing stress at a point reaches a critical value.

If $\sigma_1, \sigma_2, \sigma_3$ are the principal stresses arranged in descending order, we can write Tresca criterion as,

$$\frac{1}{2}|\sigma_1 - \sigma_2| = K_T$$

where KT is the material dependent parameter determined experimentally. If σ_y be the yield stress, the maximum shear is $\dfrac{\sigma_2}{2}$. Tresca condition can be written as,

$$|\sigma_1 - \sigma_3| = \sigma_y$$

or in terms of ρ and θ,

$$\rho \sin\left(\theta + \frac{\pi}{3}\right) = \sqrt{2k}$$

The maximum shear stress at appoint does not change when the state of stress at the point is changed hydrostatically. Tresca yield criteria represents a hexagonal cylinder in principal stress space.

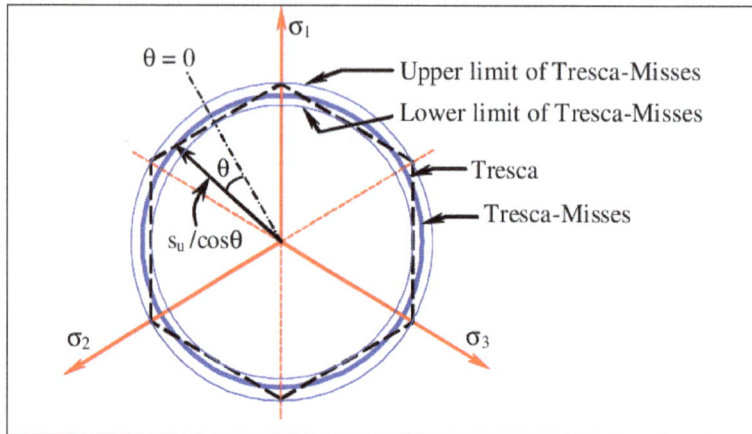

Locus of Tresca and von Mises yield
criteria on deviatoric plane.

Difference between Hot and Cold Working

Cold working may be defined as plastic deformation of metals and alloys at a temperature below the recrystallization temperature for that metal or alloy. In cold working process the strain hardening which occurs as a result of mechanical working, does not get relieved. In fact as the metal or alloys gets progressively strain hardened, more and more force is required to cause further plastic deformation. After sometime, if the effect of strain hardening is not removed, the forces applied to cause plastic deformation may cause cracking and failure of material.

Hot working may be explained as plastic deformation of metals and alloys at such a temperature above recrystallization temperature at which recovery and recrystallization take place simultaneously with the strain hardening.

Recrystallization temperature is not a fixed temperature but is actually a temperature range. Its value depends upon several factors. Some of the important factors are:

- Nature of metal or alloy: It is usually lower for pure metals and higher for alloys. For pure metals, recrystallization temperature is roughly one third of its melting point and for alloys about half of the melting temperature.

- Amount of cold work already done: The recrystallization temperature is lowered as the amount of strain-hardening done on the work piece increases.

- Strain-rate: Higher the rate of strain hardening, lower is the recrystallization temperature. For mild steel, recrystallization temperature range may be taken as 550– 650°C. Recrystallization temperature of low melting point metals like lead, zinc and tin, may be taken as room temperature. The effects of strain hardening can be removed by annealing above the recrystallization temperature.

Advantages and Disadvantages of Cold and Hot Working Processes

- As cold working is practically done at room temperature, no oxidation or tarnishing of surface takes place. No scale formation is there, hence there is no material loss where as in hot working, there is scale formation due to oxidation besides, hot working of steel also results in partial decarburization of the work piece surface as carbon gets oxidized as CO_2.

- Cold working results in better dimensional accuracy and a bright surface. Cold rolled steel bars are therefore called bright bars, while those produced by hot rolling process are called black bars (they appear greyish black due to oxidation of surface).

- In cold working heavy work hardening occurs which improves the strength and hardness of bars, and high forces are required for deformation increasing energy consumption. In hot working this is not so.

- Due to limited ductility at room temperature, production of complex shapes is not possible by cold working processes.

- Severe internal stresses are induced in the metal during cold working. If these stresses are not relieved, the component manufactured may fail prematurely in service. In hot working, there are no residual internal stresses and the mechanically worked structure is better than that produced by cold working.

- The strength of materials reduces at high temperature. Its malleability and ductility improve at high temperatures. Hence low capacity equipment is required for hot working processes. The forces on the working tools also reduce in case of hot working processes.

- Sometimes, blow holes and internal porosities are removed by welding action at high temperatures during hot working.

- Non-metallic inclusions within the work piece are broken up. Metallic and non-metallic segregations are also reduced or eliminated in hot working as diffusion is promoted at high temperatures making the composition across the entire cross-section more uniform.

Effect of Strain Rate on Forming Process

Higher the rate of strain hardening, lower is the recrystallization temperature. For mild steel, recrystallization temperature range may be taken as 550–650°C. Recrystallization temperature of low melting point metals like lead, zinc and tin, may be taken as room temperature. The effects of strain hardening can be removed by annealing above the recrystallization temperature.

Forging

Forging is a basic process in which the work piece is shaped by compressive forces applied through various dies and tooling. It is one of the oldest and most important metalworking operations used to make jewelry, coins, and various implements by hammering metal with tools made of stone. Forged parts now include large rotors for turbines; gears; bolts and rivets; cutlery; hand tools; numerous structural components for machinery, aircraft and railroads and a variety of other transportation equipment.

Simple forging operations can be performed with a heavy hammer and an anvil, as has been done traditionally by blacksmiths. However, most forgings require a set of dies and such equipment as a press or a powered forging hammer.

Forging may be carried out at room temperature (cold forging) or at elevated temperatures (warm or hot forging) depending on the homologous temperature. Cold forging requires higher forces (because of the higher strength of the work piece material), and the work piece material must possess sufficient ductility at room temperature to undergo the necessary deformation without cracking. Cold-forged parts have a good surface finish and dimensional accuracy. Hot forging requires lower forces, but the dimensional accuracy and surface finish of the parts are not as good as in cold forging. Forgings generally are subjected to additional finishing operations, such as heat treating to modify properties and machining to obtain accurate final dimensions and a good surface finish. These finishing operations can be minimized by precision forging, which is an important example of net-shape or near-net-shape forming processes. As we shall seen components that can be forged successfully also may be manufactured economically by other methods, such as casting, powder metallurgy, or machining. Each of these will produce a part having different characteristics, particularly with regard to strength, toughness, dimensional accuracy, surface finish, and the possibility of internal or external defects.

In forging the material is deformed applying either impact load or gradual load. Based on the type of loading, forging is classified as hammer forging or press forging. Hammer forging involves impact load, while press forging involves gradual loads.

Based on the nature of material flow and constraint on flow by the die/punch, forging is classified as open die forging, impression die forging and flashless forging.

Open Die Forging

In this, the work piece is compressed between two platens. There is no constraint to material flow in lateral direction. Open die forging is a process by which products are made through a series of incremental deformation using dies of relatively simple shape. The top die is attached to ram and bottom die is attached to the hammer anvil or press bed. Metal work piece is heated above recrystalline temp from 1900 to 24000 c. Most open die forging are produced on flat dies. Convex surface dies and concave surface dies are also used in pairs or with flat dies.

Open die forging is classified into three main types, namely, cogging, fullering and edging. Cogging (also called as drawing out) consists of a sequence in which the thickness of an ingot is reduced to billet or blooms by narrow dies. Fullering and Edging operations are done to reduce the cross section using convex shaped or concave shaped dies. Material gets distributed and hence gets elongated and reduction in thickness happens. Upsetting is an open die forging in which the billet is subjected to lateral flow by the flat die and punch. Due to friction the material flow across the thickness is non-uniform. Material adjacent to the die gets restrained from flowing, whereas, the material at center flows freely. This causes a phenomenon called barreling in upset forging.

Open die forging.

Impression Die Forging

Here half the impression of the finished forging is sunk or made in the top die and other half of the impression is sunk in the bottom die. In impression die forging, the work piece is pressed between the dies. As the metal spreads to fill up the cavities sunk in the dies, the requisite shape is formed between the closing dies. Some material which is forced out of the dies is called "flash". The flash provides some cushioning for the dies, as the top strikes the anvil. The flash around the work piece is cut and discarded as scrap. For a good forging, the impression in the dies has to be completely filled by the material. This may require several blows of the hammer, a single blow may not be sufficient.

Closed Die Forging

Closed die forging is very similar to impression die forging, but in true closed die forging, the amount of material initially taken is very carefully controlled, so that no flash is formed. Otherwise, the process is similar to impression die forging. It is a technique which is suitable for mass production.

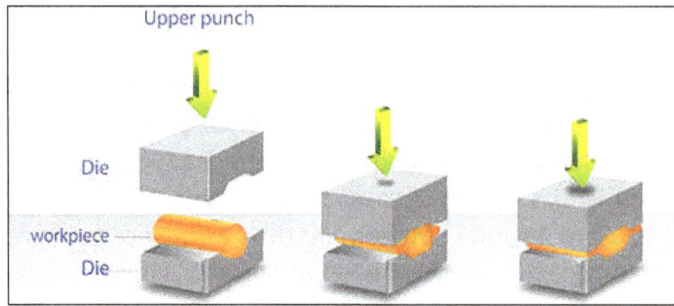

Closed die forging.

Drop Forging

Drop forging utilizes a closed impression die to obtain the desired shape of the component. The shaping is done by the repeated hammering given to the material in the die cavity. The equipment used for delivering the blows are called drop hammers.

Drop forging die consists of two halves. The lower half of the die is fixed to the anvil of the machine while the upper half is fixed to the ram. The heated stock is kept in the lower die. While the ram delivers four to five blows on the metal, in quick succession so that the metal spread and completely fills the die cavity. When the two die halves closed the complete cavity is formed.

The die impressions are machined in the die cavity, because of more complex shapes can be obtained in drop forging, compared to smith forging. However too complex shape with internal cavities, deep pockets, cannot be obtained in drop forging. Due to limitation of withdrawal of finished forging from die. The final shape desired in drop forging cannot be obtained directly from the stock in the single pass. Depending upon the shape of the component, the desired grain flow direction and the material should be manipulated in a number of passes. Various passes are used are:

1. Fullering impression: Since drop forging involves only a reduction in cross section with no upsetting, the very first step to reduce the stock is fullering. The impression machined in the die to achieve this is called fullering impression.

2. Edging impression: Also called as preform. This stage is used to gather the exact amount of material required at each cross-section of the finished component. This is the most important stage in drop forging.

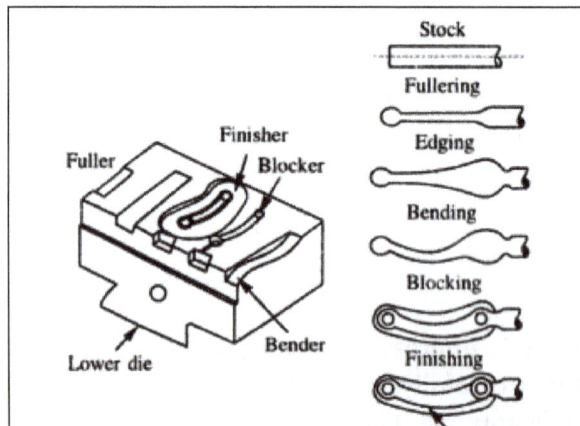

3. Bending impression: This is required for those parts which have a bend shape. The bend shape can also be obtained without the bending impressions. Then the grain flow direction will not follow the bend shape and thus the point of bend may become weak. To improve the grain flow, therefore a bending impression is incorporated after edging impression.

4. Blocking impression: It is also called as semi finshing impression. Blocking is a step before finishing. In forging, it is difficult for the material to flow to deep pockets, sharp corners etc. Hence before the actual shape is obtained, the material is allowed to have one or more blocking impressions where it requires the shape very near to final one. The blocking impression is characterized by large corner radii and fillet but no flash.

5. Finishing impression: This is the final impression where the actual shape required obtained. In order to ensure that the metal completely fills the die cavity, a little extra metal is added to the stock. The extra metal will form the flash and surround the forging in the parting plane.

6. Trimming: In this stage the extra flash present around the forging is trimmed to get the forging in the usable form.

Press Forging

Press forging dies are similar to drop forging dies as also the process. In press forging the metal is shaped not by means of a series of blows as in drop forging, but by means of a single continuous squeezing action. This squizing is obtained by means of hydraulic presses. Because of continuous action of the hydraulic presses, the material gets uniformly deformed throughout the entire depth. More hammer force is likely to be transmitted to the machine frame in drop forging where in press forging it is absorbed fully by stock. The impression obtained in press forging is clean compared to that of jarred impression which is like in drop forged component. The draft angle in press forging is less than in drop forging. But the press capacity required for deforming is higher and as a result the smaller sized component only are press forged in closed impression dies. The presses have capacities ranging from 5 MN to 50 MN for normal application For special heavy duty application, higher capacity press of order 150 MN are required.

To provide the necessary alignment the two halves, die post are attached to the bottom die so that the top die would slide only on the post and thus register the correct alignment. This ensures better tolerance for press forged components.

Forging Defects

The common forging defects can be traced to defects in raw material, improper heating of material, Faulty design of dies and improper forging practice. Most common defects present in forgings are:

1. Laps and Cracks at corners or surfaces lap is caused due to following over of a layer of material over another surface. These defects are caused by improper forging and faulty die design.

2. Incomplete forging—either due to less material or inadequate or improper flow of material.

3. Mismatched forging due to improperly aligned die halves.

4.   Scale pits—due to squeezing of scales into the metal surface during hammering action.

5.   Burnt or overheated metal—due to improper heating.

6.   Internal cracks in the forging which are caused by use of heavy hammer blows and improperly heated and soaked material.

7.   Fibre flow lines disruption due to very rapid plastic flow of metal.

Upset Forging Die Design

In upset forging there is no reduction in cross section and stock length chosen is smallest area of cross section. Here very negligible flash is provided. Depending upon the shape of upsetting ,the number of passes or blows in the dies are to be designed. The amount of upsetting done in a single stage is limited. Three rules are to be followed for safe amount of upsetting.

1. Maximum length of unsupported stock can be gathered or upset in a single pass. It is not more than three times the stock diameter. Beyond the length material is likely to buckle. Under axial upsetting load than be upset as shown below in figure.

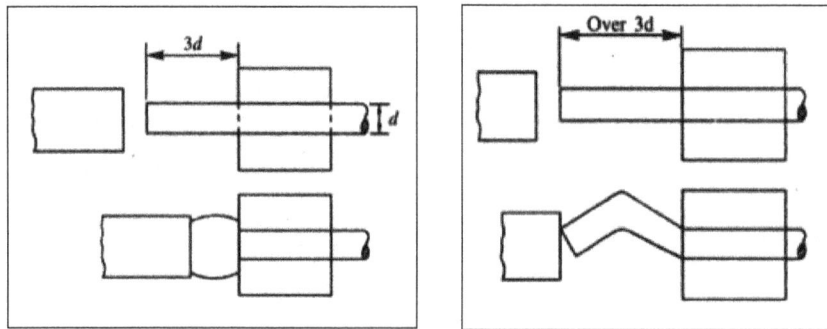

(a) Stock upset.   (b) Stock buckle.

2. If the stock longer than three times the diameter is to be upset in a single blow, then the following conditions should be complied. The die cavity should not be wider than 1.5 times the stock diameter and the free length of the stock outside the die should be less than half the stock diameter. If these conditions are not complied, the stock would bend.

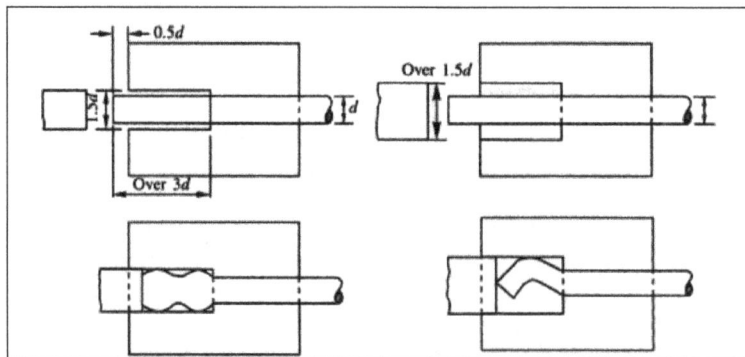

3. For upsetting the stock which is longer than three times the diameter and the free length of stock outside the die is up to 2.5 times the diameter, the following conditions should be satisfied.

The material is to be confined into a conical cavity made in punch with the mouth diameter not exceeding 1.5 times the stock diameter and bottom size being 1.25 times the stock diameter. Also the necessary that the heading tool recess be not less than two thirds the length of the work ing stock or not less than the working stock minus 2.5 times the stock diameter.

Forging Design

It is necessary to design the shape of forging to be obtained from the die.

Parting Plane

A Parting plane is the plane in which the two die halves of forging meet. It could be a simple plane ar irregularly bent, depending upon the shape of forging. The choice of proper parting plane greatly influences the cost of the die as well as the grain flow in the forging.

In any forging, the parting plane should be largest cross sectional area of forging, since it is easier to spread the metal than to force it into deep pockets. A flat parting plane is more economical. Also the parting plane chosen in such a way that the amount of material is located in each of the two die haves so that no deep die cavities are required It may be required to put more metal into the top die half since metal would flow more readily in the top half than in the bottom half.

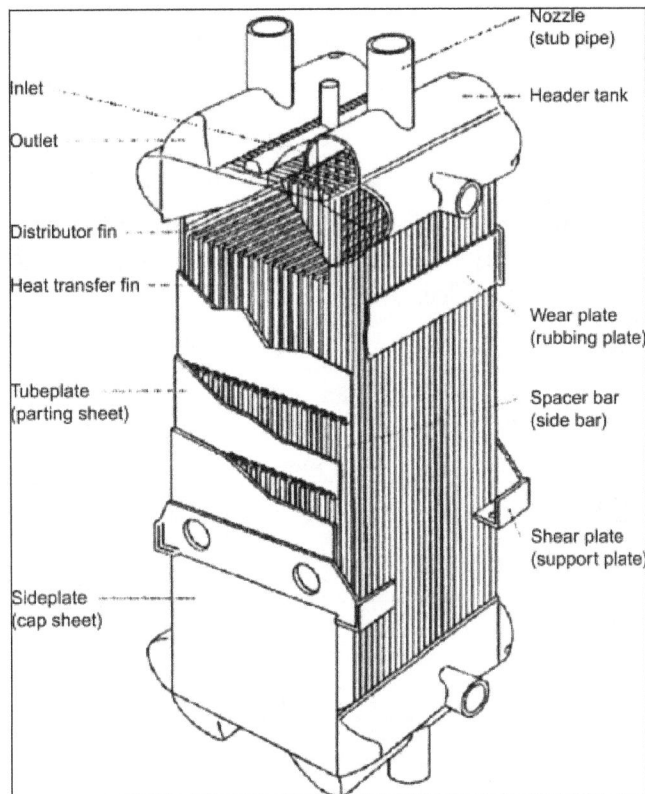

Fillet and Corner Radii

Forging involves the flow of metal in an orderly manner. Therefore it is necessary to provide a streamlined path for the flow of metal so that defects free forging are produced. When two or more surface met a corner is produced which restrict the flow of metal. Therfore these corners are to be rounded off to improve the flow of metal. Fillets are for rounding the internal angles wher corner is that of the external angles.

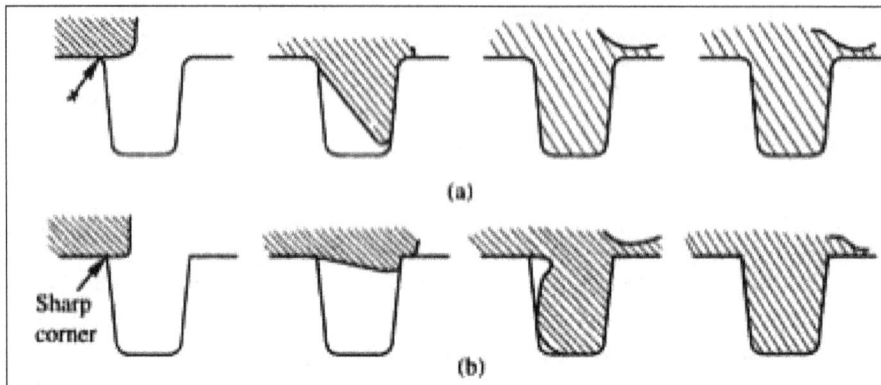

Effect of edge radius on floe metal.

Allowance

Shrinkage Allowance

The forging are generally made at room temperature of 1150 to 1300 °C.At this temperature, the material gets expanded and when it is cooled to the atmospheric temperature, it dimension would be reduced. It is very difficult to control the temperature at which forging process would be complete, therefore to precisely control the dimensions.

The forgings are generally made at a temperature of 1150 to 1300 °C. At this temperature, the material gets expanded and when it is cooled to the atmospheric temperature, its dimension would be reduced. It is difficult to control the temperature at which forging process would be complete. Therefore precisely control the dimensions.

Table: Shrinkage Allowance.

Length or width mm	Commercial + or − mm	Close +or − mm
up to 25	0.08	0.05
26 to 50	0.15	0.08
51 to 75	0.23	0.13
76 to 100	0.30	0.15
101 to125	0.38	0.20
126 to 150	0.45	0.23
Each additional 25	0.075	0.038
For example 400	1.200	0.830

Die Wear Allowance

The die wear allowance is added to account for the gradual wear of the die which takes place with the use of die.

Table: Die wear Tolerance.

Net mass of forging (kg)	Commercial + or − mm	Close + or − mm
up to 0.45	0.80	0.40
0.46 to 1.35	0.88	0.45
1.36 to 2.25	0.95	0.48
2.26 to 3.20	1.03	0.53
3.21 to 4.10	1.11	0.55
4.11 to 5.00	1.18	0.60
Each additional 1add	0.083	0.041
For example 15.00	2.010	1.010

Finish Allowance

Matching allowance is to be provided on various forged surfaces which need to be further machined. The amount of allowance to be provided should account for the accuracy, the depth of the decarburized layer. Also the scale pits that are likely to form on the component should also be removed by machining.

Table: Finish Allowance for Drop Forgings.

Greatest dimension (mm)	Minimum allowance per surface (mm)
up to 200	1.5
201 to 400	2.5
401 to 600	3.0
601 to 900	4.0
Above 900	5.0

Table: Finish Allowance for upset Forgings.

Greatest dimension (mm)	Minimum allowance per surface (mm)
up to 50	1.5
51 to 200	2.5
Above 900	3.0

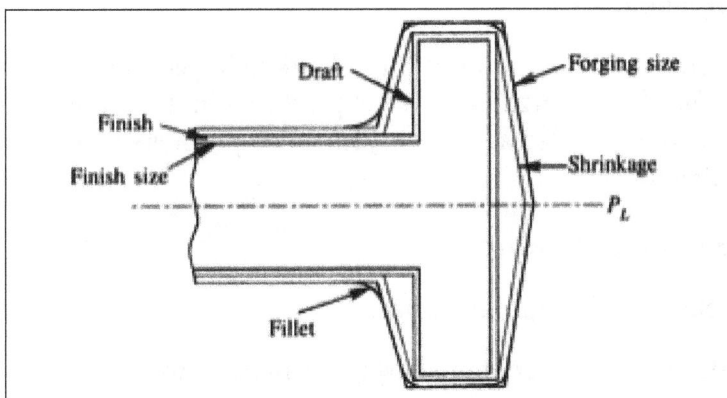

Allowances shown on forged component.

Stock

Drop forging do not get upset and therefore the stock size to be chosen depends upon the largest cross sectional area of the component. To get the stock size the flash allowance is to be provided over and above the stock volume. The stock to be used either round, rectangular or any other cross section depending upon the nature of component. Having decided on the cross section of the stock, and from the total volume of the component and the flash, it is possible to find the length of the stock. The stock of the die is to be moved from one impression to other and hence a tong hold is provided in addition to the length of the stock.

Rolling

In this process, metals and alloys are plastically deformed into semi-finished or finished products by being pressed between two rolls which are rotating. The metal is initially pushed into the space between two rolls, thereafter once the roll takes a "bite" into the edge of the material, the material gets pulled in by the friction between the surfaces of the rolls and the material. The material is subjected to high compressive force as it is squeezed (and pulled along) by the rolls. This is a process to deal with material in bulk in which the cross-section of material is reduced and its length increased. The final cross-section is determined by the impression cut in the roll surface through which the material passes and into which it is compressed.

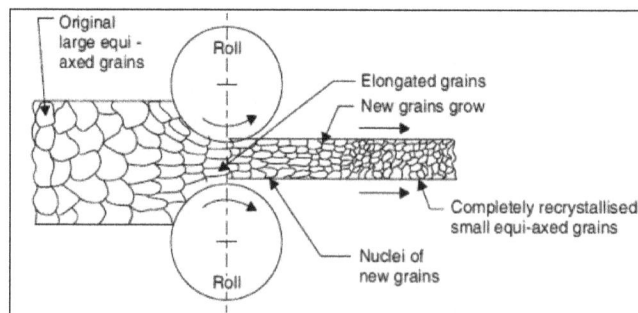

Rolling is done both hot and cold. In a rolling mill attached to a steel plant, the starting point is a cast ingot of steel which is broken down progressively into blooms, billets and slabs. The slabs are further hot rolled into plate, sheet, rod, bar, rails and other structural shapes like angles, channels etc. Conversion of steel into such commercially important sections is usually done in another rolling mill called merchant mill.

Rolling is a very convenient and economical way of producing commercially important sections. In the case of steel, about three-fourth's of all steel produced in the country is ultimately sold as a rolled product and remaining is used as forgings, extruded products and in cast form. This shows the importance of rolling process.

Nomenclature of Rolled Products

The following nomenclature is in common usage:

- Blooms: It is the first product obtained from the breakdown of Ingots. A bloom has a cross-section ranging in size from 150 mm square to 250 mm square or sometimes 250 × 300 mm rectangle.

- Billet: A billet is the next product rolled from a bloom. Billets vary from 50 mm square to 125 mm square.

- Slab: Slab is of rectangular cross-section with thickness ranging from 50 to 150 mm and is available in lengths up to 112 metres.

- Plate: A plate is generally 5 mm or thicker and is 1.0 or 1.25 metres in width and 2.5 metres in length.

- Sheet: A sheet is up to 4 mm thick and is available in same width and length as a plate.

- Flat: Flats are available in various thickness and widths and are long strips of material of specified cross-section.

- Foil: It is a very thin sheet.

- Bar: Bars are usually of circular cross-section and of several metres length. They are common stock (raw material) for capstan and turret lathes.

- Wire: A wire is a length (usually in coil form) of a small round section; the diameter of which specifies the size of the wire.

Mechanism of Rolling

Each of the two rolls contact the metal surface along the arc AB, which is called arc of contact. Arc AB divided by the radius of rolls will gives angle of contact (α). The rollers pull the material forwards only due to the friction existing between roll surface and the metal. At the moment of the bite, the reaction at the contact point A will be R acting along radial line O1A and frictional force will be acting along tangent at A at right angles to O1A. In the limiting case,

$$R \sin \alpha = \mu R \cos \alpha$$

$$\therefore \mu = \tan \alpha \ or \ \alpha = \tan{-1} \mu$$

If α is greater than tan−1 μ, the material would not enter the rolls unaided.

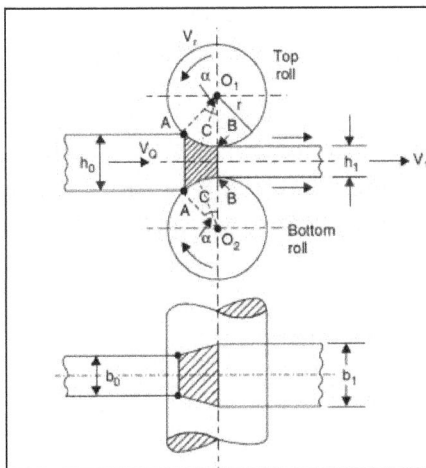

(a) Schematic diagram of rolling process.   (b) Forces during rolling.

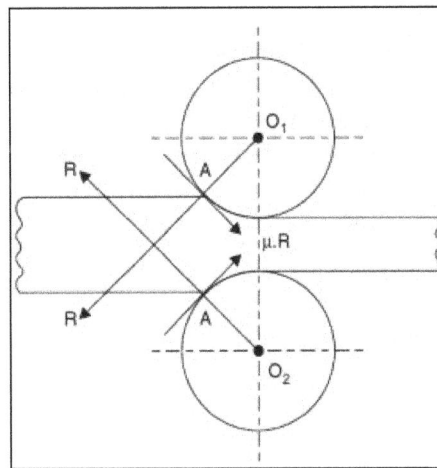

$$\cos\alpha = \frac{r - \frac{1}{2}(h_0 - h_1)}{r}$$, where ho is the thickness of material, h_1 the gap between the two rollers at the narrowest point and r is the radius of rollers. For a given diameter of rollers and gap between them the value of h_0 is limited by the value of μ which in turn depends upon the material of rolls and job being rolled, the roughness of their surfaces and the rolling temperature and speed. In case of hot rolling when maximum reduction is cross-section per pass is aimed at, it may be necessary to artificially increase the value of μ by "ragging" the surface of rolls. Ragging means making the surface of rolls rough by making fine grooves on the roll-surface. However, in cold rolling which is a finishing operation and cross-section reduction is limited, ragging of rolls is neither required nor desirable. In fact, in that case, some lubrication is resorted to in addition to giving a fine finish to the rolls. Another reason for making d_0 with a lower coefficient of friction in cold rolling is that in this process, very high pressures are used and even with a low value of μ, adequate frictional force becomes available.

The usual values of biting angles employed in industry are:

- 2–10° for cold rolling of sheets and strips.

- 15–20° for hot rolling of sheets and strips.

- 24–30° for hot rolling of heavy billets and blooms.

In the rolling process, although the material is being squeezed between two rolls, the width (b_0) of the material does not increase or increases only very slightly. Since volume of material entering the rolls is equal to the volume of material leaving the rolls, and the thickness of material reduces from h_0 to h_1, the velocity of material leaving the rolls must be higher than the velocity of material entering the rolls. The rolls are moving at a uniform r.p.m. and their surface speed remains constant. The rolls are trying to carry the material into the rolls with the help of friction alone, there is no positive grip between rolls and the material. On one side, therefore, i.e., point A where contact between the rolls and work material starts, the rolls are moving at faster surface speed than the work material. As the material gets squeezed and passes through the rollers, its speed gradually increases and at a certain section CC called neutral or no slip section, the velocity of metal equals the velocity of rolls. As material is squeezed further, its speeds exceed the speed of the rolls. The angle subtended at the centre of the roll at the neutral section is called angle of no slip or critical angle (angle BO_1C).

The deformation zone to the left of the neutral section is called lagging zone and the deformation zone to the right of the neutral section is termed leading zone. If V_r is the velocity of roll surface, V_0 the velocity of material at the entrance to the deformation zone and V_1 at the exit of the rolls, we have:

$$\text{Forward slip} = \frac{V_1 - V_r}{V_r} \times 100 \, \text{percent}$$

$$\text{Backward slip} = \frac{V_1 - V_o}{V_r} \times 100 \, \text{percent}$$

The value of forward slip normally is 3–10% and increases with increase in roll diameter and coefficient of friction and also with reduction in thickness of material being rolled. Some other useful terms associated with rolling are explained below:

$$\text{Absolute draught}: \Delta h = h_i - h_0 \text{ mm}$$

$$\text{Relative draught}: \frac{\Delta h}{h_i} \times 100 \text{ percent}$$

Absolute elongation, Δl = Final length – Original length of work material

Coefficient of elongation = Final length/Original length

Absolute spread = Final width of work material – Original width of material

Types of Rolling Mills

Different types of rolling mills are described as follows:

Two high mills: It comprises of two heavy rolls placed one over the other. The rolls are supported in bearings housed in sturdy upright frames (called stands) which are grouted to the rolling mill floor. The vertical gap between the rolls is adjustable. The rolls rotate in opposite directions and are driven by powerful electrical motors. Usually the direction of rotation of rolls cannot be altered, thus the work has to be fed into rolls from one direction only. If rolling entails more than one 'pass' in the same set of rolls, the material will have to be brought back to the same side after the first pass is over. Since transporting material (which is in red hot condition) from one side to another is difficult and time consuming (material may cool in the meantime), a "two high reversing mill" has been developed in which the direction of rotation of rolls can be changed. This facilitates rolling of material by passing it through back and forth passes. A two high rolling mill arrangement is shown in the figure below.

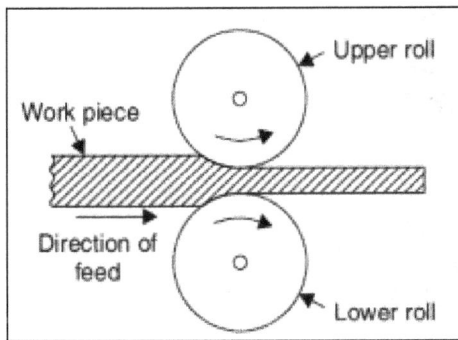

A two high mill.

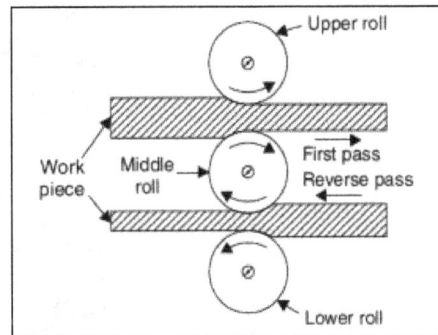

A three high rolling mill.

Three high mills: A three high rolling mill arrangement is shown in figure above. It consists of three rolls positioned directly over one another as shown. The direction of rotation of the first and second rolls are opposite as in the case of two high mill. The direction of rotation of second and third rolls is again opposite to each other. All three rolls always rotate in their bearings in the same direction. The advantage of this mill is that the work material can be fed in one direction between the first and second roll and the return pass can be provided in between the second and third rolls. This obviates the transport of material from one side of rolls to the other after one pass is over.

Four high mills: This mill consists of four horizontal rolls, two of smaller diameter and two much larger ones. The larger rolls are called backup rolls. The smaller rolls are the working rolls, but if the backup rolls were not there, due to deflection of rolls between stands, the rolled material would be thicker in the centre and thinner at either end. Backup rolls keep the working rolls pressed and restrict the deflection, when the material is being rolled. The usual products of these mills are hot and cold rolled plates and sheets.

Cluster mills: It consists of two working rolls of small diameter and four or more backing rolls. The large number of backup rolls provided becomes necessary as the backup rolls cannot exceed the diameter of working rolls by more than 2–3 times. To accommodate processes requiring high rolling loads (e.g., cold rolling of high strength steels sheets), the size of working rolls becomes small. So does the size of backup rolls and a stage may be reached that backup rolls themselvesmay offer deflection. So the backup rolls need support or backing up by further rolls.

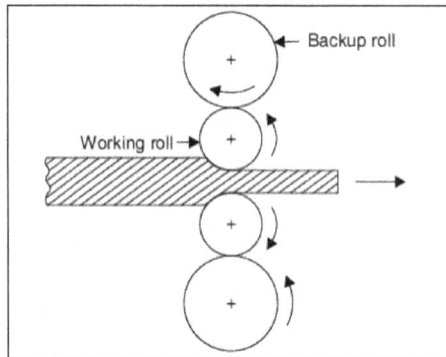

Geometric Considerations

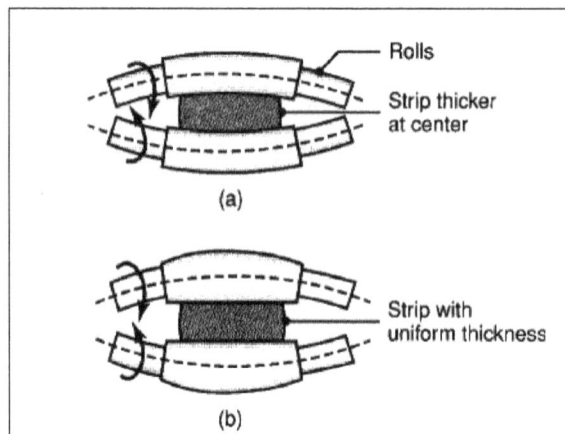

(a) Bending of straight cylindrical rolls caused by roll forces. (b) Bending of rolls ground with camber, producing a strip with uniform thickness through the strip width.

Because of the forces acting on them, rolls undergo changes in shape during rolling. just as a straight beam deflects under a transverse load, roll forces tend to bend the rolls elastically during rolling As expected, the higher the elastic modulus of the roll material, the smaller the roll deflection. As a result of roll bending, the rolled strip tends to be thicker at its center than at its edges (crown). The usual method of avoiding this problem is to grind the rolls in such way that their diameter at the center is slightly larger than at their edges (camber). Thus, when the roll bends, the strip being rolled now has a constant thickness along its width For rolling sheet metals, the radius of the maximum camber point is generally 0.25 mm greater than that at the edges of the roll. However, as expected, a particular camber is correct only for a certain load and strip width. To reduce the effects of deflection, the rolls also can be subjected to external bending by applying moments at their bearings.

Rolls and Roll Pass Design

Two types of rolls—Plain and Grooved are shown in fig. Rolls used for rolling consists of three parts viz., body, neck and wabbler. The necks rest in the bearings provided in the stands and the starshaped wabblers are connected to the driving shaft through a hollow cylinder. Wabbler acts like a safety device and saves the main body of the roll from damage if too heavy a load causes severe stresses. The actual rolling operation is performed by the body of the roll. The rolls are generally made from a special variety of cast iron, cast steel or forged steel. Plain rolls have a highly finished hard surface and are used for rolling flats, plates and sheets. Grooved rolls have grooves of various shapes cut on their periphery. One-half of the required shape of rolled product is sometimes cut in the lower roll and one-half in the upper roll, so that when the rolls are assembled into its stands, the required shape in full will be produced on the work material, once it passes (i.e., rolled) through the groove in question. However it should be understood that the desired shape of the rolled section is not achieved in a single pass. The work material has to be rolled again and again through several passes and each pass brings the cross-section of the material closer to the final shape required. These passes are carefully designed to avoid any rolling defect from creeping in. Rolling is a painstaking process as would be noticed from the scheme of passes shown in figure for conversion of a steel billet into a round bar.

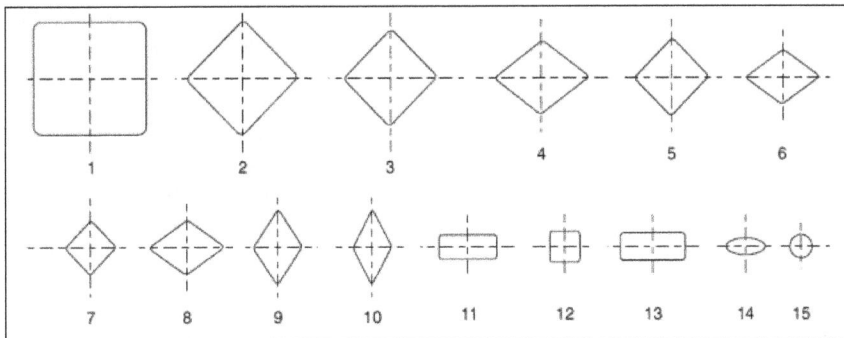

Various stages of rolling and the number of passes for converting a steel billet into round bar.

Various passes fall into the following groups:

(i) Breakdown or roughing passes,

(ii) Leader passes, and

(iii) Finishing passes.

Breakdown passes are meant to reduce the cross-sectional area. The leader passes gradually bring the cross-section of the material near the final shape. The final shape and size is achieved in finishing passes. Allowance for shrinkage on cooling is given while cutting the finishing pass grooves.

Defects in Rolling

Defects may be present on the surfaces of rolled plates and sheets, or there may be internal structural defects. Defects are undesirable not only because they compromise surface appearance, but also because they may adversely affect strength, formability, and other manufacturing characteristics. Several surface defects (such as scale, rust, scratches, gouges, pits, and cracks) have been identified in sheet metals. These defects may be caused by inclusions and impurities in the original cast material or by various other conditions related to material preparation and to the rolling operation. Wavy edges on sheets are the result of roll bending. The strip is thinner along its edges than at its center); thus, the edges elongate more than the center. Consequently, the edges buckle because they are constrained by the central region from expanding freely in the longitudinal (rolling) direction. The cracks are usually the result of poor material ductility at the rolling temperature. Because the quality of the edges of the sheet may affect sheet-metal-forming operations, edge defects in rolled sheets often are removed by shearing and slitting operations. Alligatoring is the phenomenon and typically is caused by non uniform bulk deformation of the billet during rolling or by the presence of defects in the original cast material.

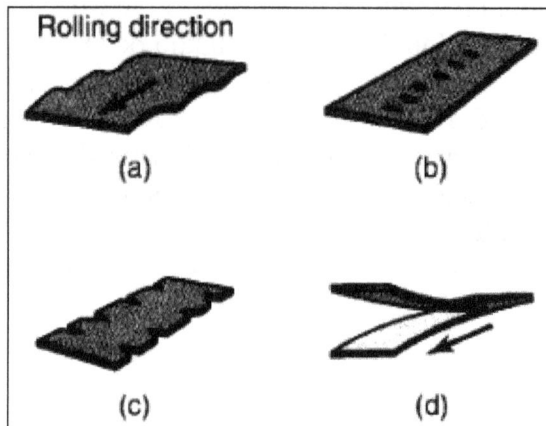

(a) wavy edges; (b) zipper cracks in the center of the strip;
(c) edge cracks; and (d) alligatoring.

EXTRUSION

Extrusion is a process used to create objects of a fixed cross-sectional profile. A material is pushed through a die of the desired cross-section. The two main advantages of this process over other manufacturing processes are its ability to create very complex cross-sections, and to work materials that are brittle, because the material only encounters compressive and shear stresses. It also forms parts with an excellent surface finish.

Drawing is a similar process, which uses the tensile strength of the material to pull it through the die. This limits the amount of change which can be performed in one step, so it is limited to simpler shapes, and multiple stages are usually needed. Drawing is the main way to produce wire. Metal bars and tubes are also often drawn.

HDPE pipe during extrusion. The HDPE material is coming from the heater,
into the die, then into the cooling tank. This Acu-Power conduit pipe is
co-extruded - black inside with a thin orange jacket, to designate power cables.

Extrusion may be continuous (theoretically producing indefinitely long material) or semi-continuous (producing many pieces). The extrusion process can be done with the material hot or cold. Commonly extruded materials include metals, polymers, ceramics, concrete, modelling clay, and foodstuffs. The products of extrusion are generally called "extrudates".

Also referred to as "hole flanging", hollow cavities within extruded material cannot be produced using a simple flat extrusion die, because there would be no way to support the centre barrier of the die. Instead, the die assumes the shape of a block with depth, beginning first with a shape profile that supports the center section. The die shape then internally changes along its length into the final shape, with the suspended center pieces supported from the back of the die. The material flows around the supports and fuses together to create the desired closed shape.

The extrusion process in metals may also increase the strength of the material.

Process

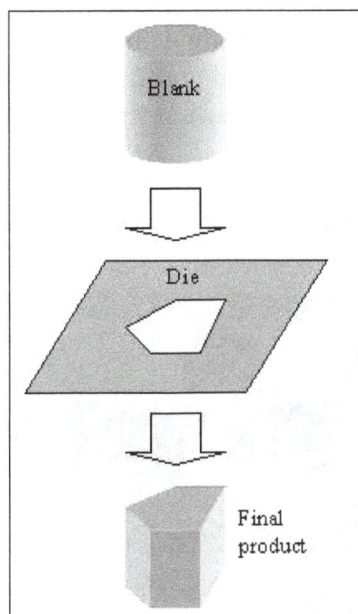

Extrusion of a round blank through a die.

The process begins by heating the stock material (for hot or warm extrusion). It is then loaded into the container in the press. A dummy block is placed behind it where the ram then presses on the material to push it out of the die. Afterward the extrusion is stretched in order to straighten it. If better properties are required then it may be heat treated or cold worked.

The extrusion ratio is defined as the starting cross-sectional area divided by the cross-sectional area of the final extrusion. One of the main advantages of the extrusion process is that this ratio can be very large while still producing quality parts.

Hot Extrusion

Hot extrusion is a hot working process, which means it is done above the material's recrystallization temperature to keep the material from work hardening and to make it easier to push the material through the die. Most hot extrusions are done on horizontal hydraulic presses that range from 230 to 11,000 metric tons (250 to 12,130 short tons). Pressures range from 30 to 700 MPa (4,400 to 101,500 psi), therefore lubrication is required, which can be oil or graphite for lower temperature extrusions, or glass powder for higher temperature extrusions. The biggest disadvantage of this process is its cost for machinery and its upkeep.

Hot extrusion temperature for various metals	
Material	Temperature [°C (°F)]
Magnesium	350–450 (650–850)
Aluminium	350–500 (650–900)
Copper	600–1100 (1200–2000)
Steel	1200–1300 (2200–2400)
Titanium	700–1200 (1300–2100)
Nickel	1000–1200 (1900–2200)
Refractory alloys	up to 2000 (4000)

The extrusion process is generally economical when producing between several kilograms (pounds) and many tons, depending on the material being extruded. There is a crossover point where roll forming becomes more economical. For instance, some steels become more economical to roll if producing more than 20,000 kg (50,000 lb).

Aluminium Hot Extrusion Die

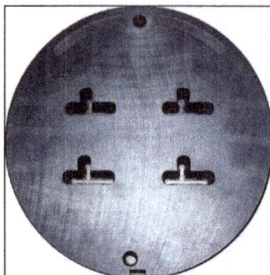

Front side of a four family die. For reference, the die is 228 mm (9.0 in) in diameter.

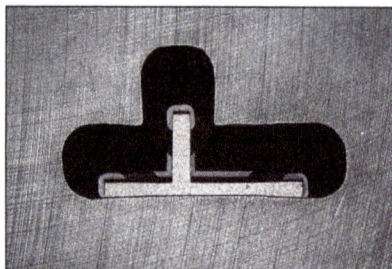

Close up of the shape cut into the die. Notice that the walls are drafted and that the back wall thickness varies.

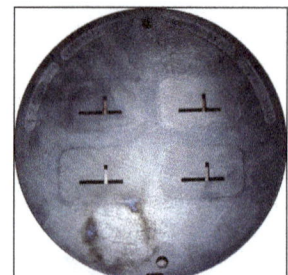

Back side of die. The wall thickness of the extrusion is 3 mm (0.12 in).

Cold Extrusion

Cold extrusion is done at room temperature or near room temperature. The advantages of this over hot extrusion are the lack of oxidation, higher strength due to cold working, closer tolerances, better surface finish, and fast extrusion speeds if the material is subject to hot shortness.

Materials that are commonly cold extruded include: lead, tin, aluminum, copper, zirconium, titanium, molybdenum, beryllium, vanadium, niobium, and steel.

Examples of products produced by this process are: collapsible tubes, fire extinguisher cases, shock absorber cylinders and gear blanks.

Warm Extrusion

Warm extrusion is done above room temperature, but below the recrystallization temperature of the material the temperatures ranges from 800 to 1800 °F (424 to 975 °C). It is usually used to achieve the proper balance of required forces, ductility and final extrusion properties.

Friction Extrusion

Friction extrusion was invented at The Welding Institute in the UK and patented in 1991. It was originally intended primarily as a method for production of homogenous microstructures and particle distributions in metal matrix composite materials. Friction extrusion differs from conventional extrusion in that the charge (billet or other precursor) rotates relative to the extrusion die. An extrusion force is applied so as to push the charge against the die. In practice either the die or the charge may rotate or they may be counter-rotating. The relative rotary motion between the charge and the die has several significant effects on the process. First, the relative motion in the plane of rotation leads to large shear stresses, hence, plastic deformation in the layer of charge in contact with and near the die. This plastic deformation is dissipated by recovery and recrystallization processes leading to substantial heating of the deforming charge. Because of the deformation heating, friction extrusion does not generally require preheating of the charge by auxiliary means potentially resulting in a more energy efficient process. Second, the substantial level of plastic deformation in the region of relative rotary motion can promote solid state welding of powders or other finely divided precursors, such as flakes and chips, effectively consolidating the charge (friction consolidation) prior to extrusion.

Microextrusion

Microextrusion is a microforming extrusion process performed at the submillimeter range. Like extrusion, metal is pushed through a die orifice, but the resulting product's cross section can fit through a 1mm square. Several microextrusion processes have been developed since microforming was envisioned in 1990. Forward (ram and billet move in the same direction) and backward (ram and billet move in the opposite direction) microextrusion were first introduced, with forward rod-backward cup and double cup extrusion methods developing later. Regardless of method, one of the greatest challenges of creating a successful microextrusion machine is the manufacture of the die and ram. "The small size of the die and ram, along with the stringent accuracy requirement, needs suitable manufacturing processes." Additionally, as Fu and Chan pointed out in a 2013 state-of-the-art technology review, several issues must still be resolved before microextrusion and other microforming technologies can be implemented more widely, including deformation load and defects, forming system stability, mechanical properties, and other size-related effects on the crystallite (grain) structure and boundaries.

Extrusion Defects

- Surface cracking: It occurs when the surface of an extrusion splits. This is often caused by the extrusion temperature, friction, or speed being too high. It can also happen at lower temperatures if the extruded product temporarily sticks to the die.

- Pipe: A flow pattern that draws the surface oxides and impurities to the center of the product. Such a pattern is often caused by high friction or cooling of the outer regions of the billet.

- Internal cracking: When the center of the extrusion develops cracks or voids. These cracks are attributed to a state of hydrostatic tensile stress at the centerline in the deformation zone in the die.

- Surface lines: When there are lines visible on the surface of the extruded profile. This depends heavily on the quality of the die production and how well the die is maintained, as some residues of the material extruded can stick to the die surface and produce the embossed lines.

Equipment

A horizontal hydraulic press for hot aluminum extrusion (loose dies and scrap visible in foreground).

There are many different variations of extrusion equipment. They vary by four major characteristics:

- Movement of the extrusion with relation to the ram. If the die is held stationary and the ram moves towards it then it is called "direct extrusion". If the ram is held stationary and the die moves towards the ram it is called "indirect extrusion".

- The position of the press, either vertical or horizontal.

- The type of drive, either hydraulic or mechanical.

- The type of load applied, either conventional (variable) or hydrostatic.

A single or twin screw auger, powered by an electric motor, or a ram, driven by hydraulic pressure (often used for steel and titanium alloys), oil pressure (for aluminium), or in other specialized processes such as rollers inside a perforated drum for the production of many simultaneous streams of material.

Typical extrusion presses cost more than $100,000, whereas dies can cost up to $2000.

Forming Internal Cavities

There are several methods for forming internal cavities in extrusions. One way is to use a hollow billet and then use a fixed or floating mandrel. A fixed mandrel, also known as a German type, means it is integrated into the dummy block and stem. A floating mandrel, also known as a French type, floats in slots in the dummy block and aligns itself in the die when extruding. If a solid billet is used as the feed material then it must first be pierced by the mandrel before extruding through the die. A special press is used in order to control the mandrel independently from the ram. The solid billet could also be used with a spider die, porthole die or bridge die. All of these types of dies incorporate the mandrel in the die and have "legs" that hold the mandrel in place. During extrusion the metal divides, flows around the legs, then merges, leaving weld lines in the final product.

Two-piece aluminum extrusion die set (parts shown separated.) The male part (at right) is for forming the internal cavity in the resulting round tube extrusion.

Direct Extrusion

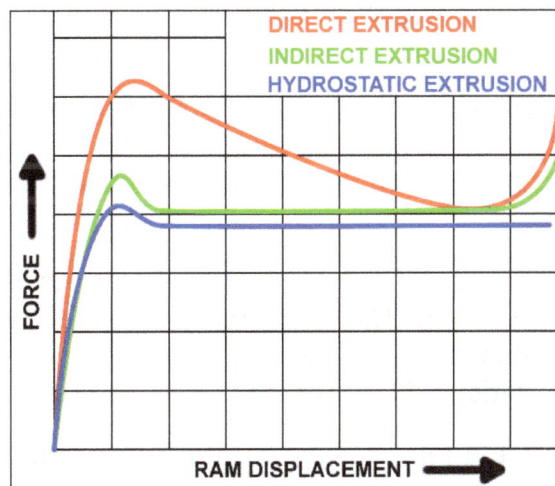

Plot of forces required by various extrusion processes.

Direct extrusion, also known as forward extrusion, is the most common extrusion process. It works by placing the billet in a heavy walled container. The billet is pushed through the die by a ram or screw. There is a reusable dummy block between the ram and the billet to keep them separated. The major disadvantage of this process is that the force required to extrude the billet is greater than that needed in the indirect extrusion process because of the frictional forces introduced by the need for the billet to travel the entire length of the container. Because of this the greatest force required is at the beginning of process and slowly decreases as the billet is used up. At the end of the billet the force greatly increases because the billet is thin and the material must flow radially to exit the die. The end of the billet (called the butt end) is not used for this reason.

Indirect Extrusion

In indirect extrusion, also known as backwards extrusion, the billet and container move together while the die is stationary. The die is held in place by a "stem" which has to be longer than the container length. The maximum length of the extrusion is ultimately dictated by the column strength of the stem. Because the billet moves with the container the frictional forces are eliminated. This leads to the following advantages:

- A 25 to 30% reduction of friction, which allows for extruding larger billets, increasing speed, and an increased ability to extrude smaller cross-sections.

- There is less of a tendency for extrusions to crack because there is no heat formed from friction.

- The container liner will last longer due to less wear.

- The billet is used more uniformly so extrusion defects and coarse grained peripherals zones are less likely.

The disadvantages are:

- Impurities and defects on the surface of the billet affect the surface of the extrusion. These defects ruin the piece if it needs to be anodized or the aesthetics are important. In order to get around this the billets may be wire brushed, machined or chemically cleaned before being used.

- This process isn't as versatile as direct extrusions because the cross-sectional area is limited by the maximum size of the stem.

Hydrostatic Extrusion

In the hydrostatic extrusion process the billet is completely surrounded by a pressurized liquid, except where the billet contacts the die. This process can be done hot, warm, or cold, however the temperature is limited by the stability of the fluid used. The process must be carried out in a sealed cylinder to contain the hydrostatic medium. The fluid can be pressurized two ways:

- *Constant-rate extrusion*: A ram or plunger is used to pressurize the fluid inside the container.

- *Constant-pressure extrusion*: A pump is used, possibly with a pressure intensifier, to pressurize the fluid, which is then pumped to the container.

The advantages of this process include:

- No friction between the container and the billet reduces force requirements. This ultimately allows for faster speeds, higher reduction ratios, and lower billet temperatures.

- Usually the ductility of the material increases when high pressures are applied.

- An even flow of material.

- Large billets and large cross-sections can be extruded.

- No billet residue is left on the container walls.

The disadvantages are:

- The billets must be prepared by tapering one end to match the die entry angle. This is needed to form a seal at the beginning of the cycle. Usually the entire billet needs to be machined to remove any surface defects.

- Containing the fluid under high pressures can be difficult.

- A billet remnant or a plug of a tougher material must be left at the end of the extrusion to prevent a sudden release of the extrusion fluid.

Drives

Most modern direct or indirect extrusion presses are hydraulically driven, but there are some small mechanical presses still used. Of the hydraulic presses there are two types: direct-drive oil presses and accumulator water drives.

Direct-drive oil presses are the most common because they are reliable and robust. They can deliver over 35 MPa (5000 psi). They supply a constant pressure throughout the whole billet. The disadvantage is that they are slow, between 50 and 200 mm/s (2–8 ips).

Accumulator water drives are more expensive and larger than direct-drive oil presses, and they lose about 10% of their pressure over the stroke, but they are much faster, up to 380 mm/s (15 ips). Because of this they are used when extruding steel. They are also used on materials that must be heated to very hot temperatures for safety reasons.

Hydrostatic extrusion presses usually use castor oil at pressure up to 1400 MPa (200 ksi). Castor oil is used because it has good lubricity and high pressure properties.

Die Design

The design of an extrusion profile has a large impact on how readily it can be extruded. The maximum size for an extrusion is determined by finding the smallest circle that will fit around the cross-section, this is called the *circumscribing circle*. This diameter, in turn, controls the size of the die required, which ultimately determines if the part will fit in a given press. For example, a larger press can handle 60 cm (24 in) diameter circumscribing circles for aluminium and 55 cm (22 in) diameter circles for steel and titanium.

The complexity of an extruded profile can be roughly quantified by calculating the *shape factor*, which is the amount of surface area generated per unit mass of extrusion. This affects the cost of tooling as well as the rate of production.

Thicker sections generally need an increased section size. In order for the material to flow properly legs should not be more than ten times longer than their thickness. If the cross-section is asymmetrical, adjacent sections should be as close to the same size as possible. Sharp corners should be avoided; for aluminium and magnesium the minimum radius should be 0.4 mm (1/64 in) and for steel corners should be 0.75 mm (0.030 in) and fillets should be 3 mm (0.12 in). The following table lists the minimum cross-section and thickness for various materials.

Material	Minimum cross-section [cm² (sq. in.)]	Minimum thickness [mm (in.)]
Carbon steels	2.5 (0.40)	3.00 (0.120)
Stainless steel	3.0–4.5 (0.45–0.70)	3.00–4.75 (0.120–0.187)
Titanium	3.0 (0.50)	3.80 (0.150)
Aluminium	< 2.5 (0.40)	0.70 (0.028)
Magnesium	< 2.5 (0.40)	1.00 (0.040)

Materials

Metal

Metals that are commonly extruded include:

- Aluminium is the most commonly extruded material. Aluminium can be hot or cold extruded. If it is hot extruded it is heated to 575 to 1100 °F (300 to 600 °C). Examples of products include profiles for tracks, frames, rails, mullions, and heat sinks.

- Brass is used to extrude corrosion free rods, automobile parts, pipe fittings, engineering parts.

- Copper (1100 to 1825 °F (600 to 1000 °C)) pipe, wire, rods, bars, tubes, and welding electrodes. Often more than 100 ksi (690 MPa) is required to extrude copper.

- Lead and tin (maximum 575 °F (300 °C)) pipes, wire, tubes, and cable sheathing. Molten lead may also be used in place of billets on vertical extrusion presses.

- Magnesium (575 to 1100 °F (300 to 600 °C)) aircraft parts and nuclear industry parts. Magnesium is about as extrudable as aluminum.

- Zinc (400 to 650 °F (200 to 350 °C)) rods, bar, tubes, hardware components, fitting, and handrails.

- Steel (1825 to 2375 °F (1000 to 1300 °C)) rods and tracks. Usually plain carbon steel is extruded, but alloy steel and stainless steel can also be extruded.

- Titanium (1100 to 1825 °F (600 to 1000 °C)) aircraft components including seat tracks, engine rings, and other structural parts.

Magnesium and aluminium alloys usually have a 0.75 µm (30 µin) RMS or better surface finish. Titanium and steel can achieve a 3 micrometres (120 µin) RMS.

In 1950, Ugine Séjournet, of France, invented a process which uses glass as a lubricant for extruding steel. The Ugine-Sejournet, or Sejournet, process is now used for other materials that have melting temperatures higher than steel or that require a narrow range of temperatures to extrude, such as the platinum-iridium alloy used to make kilogram mass standards. The process starts by heating the materials to the extruding temperature and then rolling it in glass powder. The glass melts and forms a thin film, 20 to 30 mils (0.5 to 0.75 mm), in order to separate it from chamber walls and allow it to act as a lubricant. A thick solid glass ring that is 0.25 to 0.75 in (6 to 18 mm) thick is placed in the chamber on the die to lubricate the extrusion as it is forced through the die. A second advantage of this glass ring is its ability to insulate the heat of the billet from the die. The extrusion will have a 1 mil thick layer of glass, which can be easily removed once it cools.

Another breakthrough in lubrication is the use of phosphate coatings. With this process, in conjunction with glass lubrication, steel can be cold extruded. The phosphate coat absorbs the liquid glass to offer even better lubricating properties.

Plastic

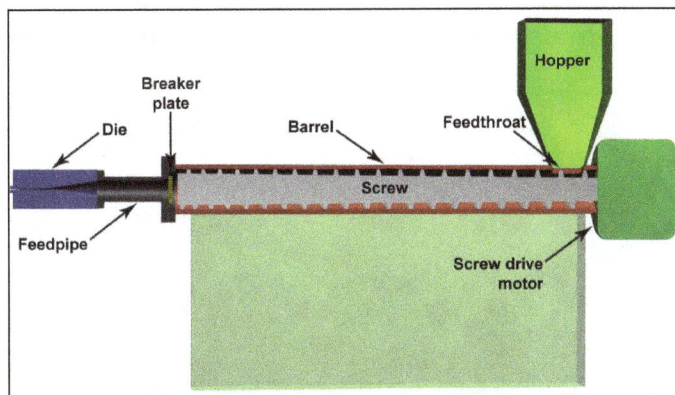

Sectional view of a plastic extruder showing the components.

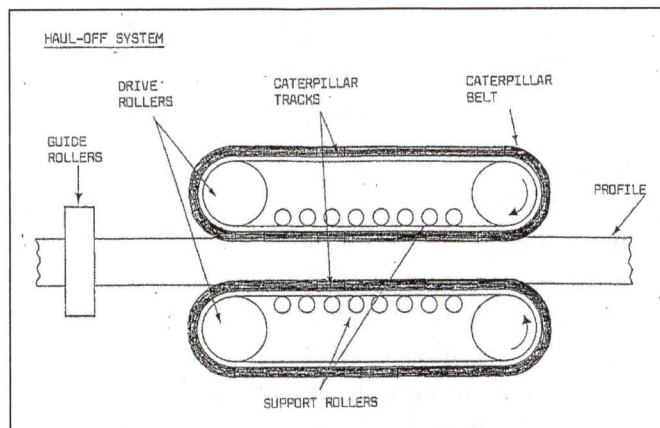

Sectional view of how a caterpillar haul-off
provides line tension.

Plastics extrusion commonly uses plastic chips or pellets, which are usually dried, to drive out moisture, in a hopper before going to the feed screw. The polymer resin is heated to molten state by a combination of heating elements and shear heating from the extrusion screw. The screw, or screws as the case with twin screw extrusion, forces the resin through a die, forming the resin into the desired shape. The extrudate is cooled and solidified as it is pulled through the die or water tank. A "caterpillar haul-off" (called a "puller") is used to provide tension on the extrusion line which is essential for overall quality of the extrudate. Pelletizers can also create this tension while pulling extruded strands in to be cut. The caterpillar haul-off must provide a consistent pull; otherwise, variation in cut lengths or distorted product will result. In some cases (such as fibre-reinforced tubes) the extrudate is pulled through a very long die, in a process called "pultrusion". The configuration of the interior screws are a driving force dependent on the application. Mixing elements or convey elements are used in various formations. Extrusion is common in the application of adding colorant to molten plastic thus creating specific custom color.

A multitude of polymers are used in the production of plastic tubing, pipes, rods, rails, seals, and sheets or films.

Ceramic

Ceramic can also be formed into shapes via extrusion. Terracotta extrusion is used to produce pipes. Many modern bricks are also manufactured using a brick extrusion process.

Applications

Food

Elbow macaroni is an extruded hollow pasta.

With the advent of industrial manufacturing, extrusion found application in food processing of instant foods and snacks, along with its already known uses in plastics and metal fabrication. The main role of extrusion was originally developed for conveying and shaping fluid forms of processed raw materials. Present day, extrusion cooking technologies and capabilities have developed into sophisticated processing functions including: mixing, conveying, shearing, separation, heating, cooling, shaping, co-extrusion, venting volatiles and moisture, encapsulation, flavor generation and sterilization. Products such as certain pastas, many breakfast cereals, premade cookie dough, some french fries, certain baby foods, dry or semi-moist pet food and ready-to-eat snacks are mostly manufactured by extrusion. It is also used to produce modified starch, and to pelletize animal feed.

Generally, high-temperature extrusion is used for the manufacture of ready-to-eat snacks, while cold extrusion is used for the manufacture of pasta and related products intended for later cooking and consumption. The processed products have low moisture and hence considerably higher shelf life, and provide variety and convenience to consumers.

In the extrusion process, raw materials are first ground to the correct particle size. The dry mix is passed through a pre-conditioner, in which other ingredients may be added, and steam is injected to start the cooking process. The preconditioned mix is then passed through an extruder, where it is forced through a die and cut to the desired length. The cooking process takes place within the extruder where the product produces its own friction and heat due to the pressure generated (10–20 bar). The main independent parameters during extrusion cooking are feed rate, particle size of the raw material, barrel temperature, screw speed and moisture content. The extruding process can induce both protein denaturation and starch gelatinization, depending on inputs and parameters. Sometimes, a catalyst is used, for example, when producing texturised vegetable proteins (TVP).

Drug Carriers

For use in pharmaceutical products, extrusion through nano-porous, polymeric filters is being used to produce suspensions of lipid vesicles liposomes or transfersomes with a particular size of a narrow size distribution. The anti-cancer drug Doxorubicin in liposome delivery system is formulated by extrusion, for example. Hot melt extrusion is also utilized in pharmaceutical solid oral dose processing to enable delivery of drugs with poor solubility and bioavailability. Hot melt extrusion has been shown to molecularly disperse poorly soluble drugs in a polymer carrier increasing dissolution rates and bioavailability. The process involves the application of heat, pressure and agitation to mix materials together and 'extrude' them through a die. Twin-screw high shear extruders blend materials and simultaneously break up particles. The resulting particle can be blended with compression aids and compressed into tablets or filled into unit dose capsules.

Biomass Briquettes

The extrusion production technology of fuel briquettes is the process of extrusion screw wastes (straw, sunflower husks, buckwheat, etc.) or finely shredded wood waste (sawdust) under high pressure when heated from 160 to 350°C. The resulting fuel briquettes do not include any of the binders, but one natural – the lignin contained in the cells of plant wastes. The temperature during compression causes melting of the surface of bricks, making it more solid, which is important for the transportation of briquettes.

MOLDING

Molding is the process of manufacturing by shaping liquid or pliable raw material us-ing a rigid frame called a mold or matrix. This itself may have been made using a pattern or model of the final object.

A mold is a hollowed-out block that is filled with a liquid or pliable material like plastic, glass, metal, or ceramic raw materials. The liquid hardens or sets inside the mold, adopting its shape. A mold is the counterpart to a cast. The very common bi-valve molding process uses two molds, one for each half of the object. Piece-molding uses a number of different molds, each creating a section of a complicated object. This is generally only used for larger and more valuable objects.

The manufacturer who makes the molds is called the mold-maker. A release agent is typically used to make removal of the hardened/set substance from the mold easier. Typical uses for molded plastics include molded furniture, molded household goods, molded cases, and structural materials.

Types of molding include:

- Blow molding.
- Powder metallurgy plus sintering.
- Compression molding.
- Extrusion molding.
- Injection molding.
- Laminating.
- Reaction injection molding.
- Matrix molding.
- Rotational molding (or Rotomolding).
- Spin casting.
- Transfer molding.
- Thermoforming.

Injection Molding

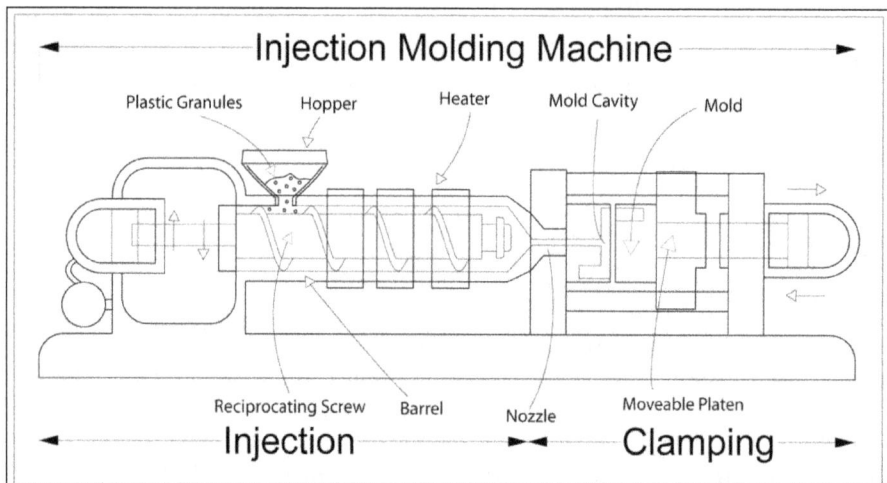

An injection molding machine.

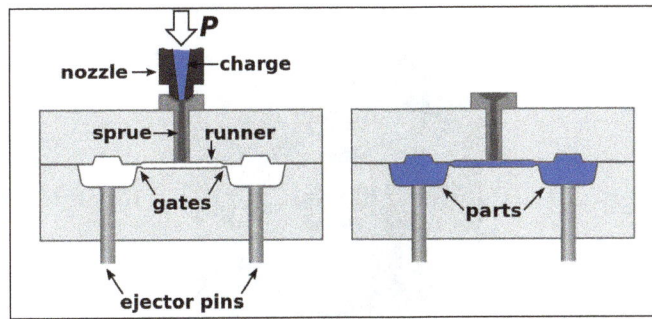

Simplified diagram of the process.

Injection molding is a manufacturing process for producing parts by injecting molten material into a mold. Injection molding can be performed with a host of materials mainly including metals (for which the process is called die-casting), glasses, elastomers, confections, and most commonly thermoplastic and thermosetting polymers. Material for the part is fed into a heated barrel, mixed (Using a helical shaped screw), and injected (Forced) into a mold cavity, where it cools and hardens to the configuration of the cavity. After a product is designed, usually by an industrial designer or an engineer, molds are made by a mold-maker (or toolmaker) from metal, usually either steel or aluminium, and precision-machined to form the features of the desired part. Injection molding is widely used for manufacturing a variety of parts, from the smallest components to entire body panels of cars. Advances in 3D printing technology, using photopolymers which do not melt during the injection molding of some lower temperature thermoplastics, can be used for some simple injection molds.

Parts to be injection molded must be very carefully designed to facilitate the molding process; the material used for the part, the desired shape and features of the part, the material of the mold, and the properties of the molding machine must all be taken into account. The versatility of injection molding is facilitated by this breadth of design considerations and possibilities.

Applications

Injection molding is used to create many things such as wire spools, packaging, bottle caps, automotive parts and components, toys, pocket combs, some musical instruments (and parts of them), one-piece chairs and small tables, storage containers, mechanical parts (including gears), and most other plastic products available today. Injection molding is the most common modern method of manufacturing plastic parts; it is ideal for producing high volumes of the same object.

Process Characteristics

Injection molding uses a ram or screw-type plunger to force molten plastic material into a mold cavity; this solidifies into a shape that has conformed to the contour of the mold. It is most commonly used to process both thermoplastic and thermosetting polymers, with the volume used of the former being considerably higher. Thermoplastics are prevalent due to characteristics which make them highly suitable for injection molding, such as the ease with which they may be recycled, their versatility allowing them to be used in a wide variety of applications, and their ability to soften and flow upon heating. Thermoplastics also have an element of safety over thermosets; if a thermosetting polymer is not ejected from the injection barrel in a timely manner, chemical crosslinking may occur causing the screw and check valves to seize and potentially damaging the injection molding machine.

Thermoplastic resin pellets for injection molding.

Injection molding consists of the high pressure injection of the raw material into a mold which shapes the polymer into the desired shape. Molds can be of a single cavity or multiple cavities. In multiple cavity molds, each cavity can be identical and form the same parts or can be unique and form multiple different geometries during a single cycle. Molds are generally made from tool steels, but stainless steels and aluminium molds are suitable for certain applications. Aluminium molds are typically ill-suited for high volume production or parts with narrow dimensional tolerances, as they have inferior mechanical properties and are more prone to wear, damage, and deformation during the injection and clamping cycles; however, aluminium molds are cost-effective in low-volume applications, as mold fabrication costs and time are considerably reduced. Many steel molds are designed to process well over a million parts during their lifetime and can cost hundreds of thousands of dollars to fabricate.

When thermoplastics are molded, typically pelletised raw material is fed through a hopper into a heated barrel with a reciprocating screw. Upon entrance to the barrel, the temperature increases and the Van der Waals forces that resist relative flow of individual chains are weakened as a result of increased space between molecules at higher thermal energy states. This process reduces its viscosity, which enables the polymer to flow with the driving force of the injection unit. The screw delivers the raw material forward, mixes and homogenises the thermal and viscous distributions of the polymer, and reduces the required heating time by mechanically shearing the material and adding a significant amount of frictional heating to the polymer. The material feeds forward through a check valve and collects at the front of the screw into a volume known as a shot.

A shot is the volume of material that is used to fill the mold cavity, compensate for shrinkage, and provide a cushion (approximately 10% of the total shot volume, which remains in the barrel and prevents the screw from bottoming out) to transfer pressure from the screw to the mold cavity. When enough material has gathered, the material is forced at high pressure and velocity into the part forming cavity. The exact amount of shrinkage is a function of the resin being used, and can be relatively predictable. To prevent spikes in pressure, the process normally uses a transfer position corresponding to a 95–98% full cavity where the screw shifts from a constant velocity to a constant pressure control. Often injection times are well under 1 second. Once the screw reaches the transfer position the packing pressure is applied, which completes mold filling and compensates for thermal shrinkage, which is quite high for thermoplastics relative to many other materials. The packing pressure is applied until the gate (cavity entrance) solidifies. Due to its small size, the gate is normally the first place to solidify through its entire thickness. Once the gate solidifies, no more material can enter the cavity; accordingly, the screw reciprocates and acquires material for the next cycle while the material within the mold cools so that it can be ejected and be dimensionally stable. This cooling duration is dramatically reduced by the use of cooling lines circulating water or oil from an external temperature controller. Once the required temperature has been achieved, the mold opens and an array of pins, sleeves, strippers, etc. are driven forward to demold the article. Then, the mold closes and the process is repeated.

For a two shot mold, two separate materials are incorporated into one part. This type of injection molding is used to add a soft touch to knobs, to give a product multiple colours, or to produce a part with multiple performance characteristics.

For thermosets, typically two different chemical components are injected into the barrel. These components immediately begin irreversible chemical reactions which eventually crosslinks the material into a single connected network of molecules. As the chemical reaction occurs, the two fluid components permanently transform into a viscoelastic solid. Solidification in the injection barrel and screw can be problematic and have financial repercussions; therefore, minimising the thermoset curing within the barrel is vital. This typically means that the residence time and temperature of the chemical precursors are minimised in the injection unit. The residence time can be reduced by minimising the barrel's volume capacity and by maximising the cycle times. These factors have led to the use of a thermally isolated, cold injection unit that injects the reacting chemicals into a thermally isolated hot mold, which increases the rate of chemical reactions and results in shorter time required to achieve a solidified thermoset component. After the part has solidified, valves close to isolate the injection system and chemical precursors, and the mold opens to eject the molded parts. Then, the mold closes and the process repeats.

Pre-molded or machined components can be inserted into the cavity while the mold is open, allowing the material injected in the next cycle to form and solidify around them. This process is known as Insert molding and allows single parts to contain multiple materials. This process is often used to create plastic parts with protruding metal screws, allowing them to be fastened and unfastened repeatedly. This technique can also be used for In-mold labelling and film lids may also be attached to molded plastic containers.

A parting line, sprue, gate marks, and ejector pin marks are usually present on the final part. None of these features are typically desired, but are unavoidable due to the nature of the process. Gate marks occur at the gate which joins the melt-delivery channels (sprue and runner) to the part forming cavity. Parting line and ejector pin marks result from minute misalignments, wear, gaseous vents, clearances for adjacent parts in relative motion, and/or dimensional differences of the mating surfaces contacting the injected polymer. Dimensional differences can be attributed to non-uniform, pressure-induced deformation during injection, machining tolerances, and non-uniform thermal expansion and contraction of mold components, which experience rapid cycling during the injection, packing, cooling, and ejection phases of the process. Mold components are often designed with materials of various coefficients of thermal expansion. These factors cannot be simultaneously accounted for without astronomical increases in the cost of design, fabrication, processing, and quality monitoring. The skillful mold and part designer will position these aesthetic detriments in hidden areas if feasible.

Examples of Polymers Best Suited for the Process

Most polymers, sometimes referred to as resins, may be used, including all thermoplastics, some thermosets, and some elastomers. Since 1995, the total number of available materials for injection molding has increased at a rate of 750 per year; there were approximately 18,000 materials available when that trend began. Available materials include alloys or blends of previously developed materials, so product designers can choose the material with the best set of properties from a vast selection. Major criteria for selection of a material are the strength and function required for the final part, as well as the cost, but also each material has different parameters for molding that must be taken into account. Other considerations when choosing an injection molding material include flexural modulus of elasticity, or the degree to which a material can be bent without damage, as well as heat deflection and water absorption. Common polymers like epoxy and phenolic are examples of thermosetting plastics while nylon, polyethylene, and polystyrene are thermoplastic. Until comparatively recently, plastic springs were not possible, but advances in polymer properties make them now quite practical. Applications include buckles for anchoring and disconnecting the outdoor-equipment webbing.

Equipment

Paper clip mold opened in molding machine; the nozzle is visible at right.

Injection molding machines consist of a material hopper, an injection ram or screw-type plunger, and a heating unit. Also known as platens, they hold the molds in which the components are shaped. Presses are rated by tonnage, which expresses the amount of clamping force that the machine can exert. This force keeps the mold closed during the injection process. Tonnage can vary from less than 5 tons to over 9,000 tons, with the higher figures used in comparatively few manufacturing operations. The total clamp force needed is determined by the projected area of the part being molded. This projected area is multiplied by a clamp force of from 1.8 to 7.2 tons for each square centimetre of the projected areas. As a rule of thumb, 4 or 5 tons/in² can be used for most products. If the plastic material is very stiff, it will require more injection pressure to fill the mold, and thus more clamp tonnage to hold the mold closed. The required force can also be determined by the material used and the size of the part. Larger parts require higher clamping force.

Mold

Mold or die are the common terms used to describe the tool used to produce plastic parts in molding.

Since molds have been expensive to manufacture, they were usually only used in mass production where thousands of parts were being produced. Typical molds are constructed from hardened steel, pre-hardened steel, aluminium, and/or beryllium-copper alloy. The choice of material to build a mold from is primarily one of economics; in general, steel molds cost more to construct, but their longer lifespan will offset the higher initial cost over a higher number of parts made before wearing out. Pre-hardened steel molds are less wear-resistant and are used for lower volume requirements or larger components; their typical steel hardness is 38–45 on the Rockwell-C scale. Hardened steel molds are heat treated after machining; these are by far superior in terms of wear resistance and lifespan. Typical hardness ranges between 50 and 60 Rockwell-C (HRC). Aluminium molds can cost substantially less, and when designed and machined with modern computerised equipment can be economical for molding tens or even hundreds of thousands of parts. Beryllium copper is used in areas of the mold that require fast heat removal or areas that see the most shear heat generated. The molds can be manufactured either by CNC machining or by using electrical discharge machining processes.

Injection Molding Die with Side Pulls

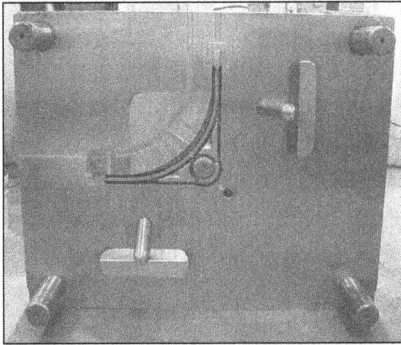

"A" side of die for 25% glass-filled acetal with 2 side pulls.

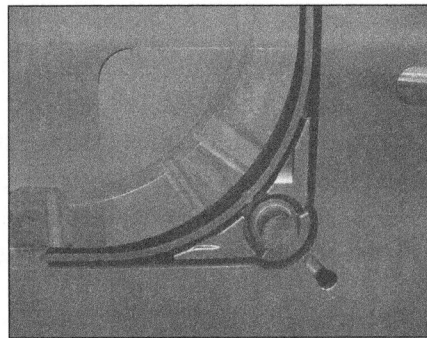

Close up of removable insert in "A" side.

"B" side of die with side pull actuators.

Insert removed from die.

Mold Design

The mold consists of two primary components, the injection mold (A plate) and the ejector mold (B plate). These components are also referred to as *molder* and *moldmaker*. Plastic resin enters the mold through a *sprue* or *gate* in the injection mold; the sprue bushing is to seal tightly against the nozzle of the injection barrel of the molding machine and to allow molten plastic to flow from the barrel into the mold, also known as the *cavity*. The sprue bushing directs the molten plastic to the cavity images through channels that are machined into the faces of the A and B plates. These channels allow plastic to run along them, so they are referred to as *runners*. The molten plastic flows through the runner and enters one or more specialised gates and into the cavity geometry to form the desired part.

Standard two plates tooling – core and cavity are inserts
in a mold base – "family mold" of five different parts.

The amount of resin required to fill the sprue, runner and cavities of a mold comprises a "shot". Trapped air in the mold can escape through air vents that are ground into the parting line of the mold, or around ejector pins and slides that are slightly smaller than the holes retaining them. If the trapped air is not allowed to escape, it is compressed by the pressure of the incoming material and squeezed into the corners of the cavity, where it prevents filling and can also cause other defects. The air can even become so compressed that it ignites and burns the surrounding plastic material.

To allow for removal of the molded part from the mold, the mold features must not overhang one another in the direction that the mold opens, unless parts of the mold are designed to move from between such overhangs when the mold opens (using components called Lifters).

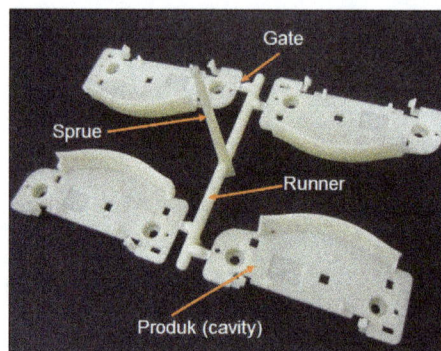

Sprue, runner and gates in actual injection molding product.

Sides of the part that appear parallel with the direction of draw (the axis of the cored position (hole) or insert is parallel to the up and down movement of the mold as it opens and closes) are typically angled slightly, called draft, to ease release of the part from the mold. Insufficient draft can cause deformation or damage. The draft required for mold release is primarily dependent on the depth of the cavity; the deeper the cavity, the more draft necessary. Shrinkage must also be taken into account when determining the draft required. If the skin is too thin, then the molded part will tend to shrink onto the cores that form while cooling and cling to those cores, or the part may warp, twist, blister or crack when the cavity is pulled away.

A mold is usually designed so that the molded part reliably remains on the ejector (B) side of the mold when it opens, and draws the runner and the sprue out of the (A) side along with the parts. The part then falls freely when ejected from the (B) side. Tunnel gates, also known as submarine or mold gates, are located below the parting line or mold surface. An opening is machined into the surface of the mold on the parting line. The molded part is cut (by the mold) from the runner system on ejection from the mold. Ejector pins, also known as knockout pins, are circular pins placed in either half of the mold (usually the ejector half), which push the finished molded product, or runner system out of a mold. The ejection of the article using pins, sleeves, strippers, etc., may cause undesirable impressions or distortion, so care must be taken when designing the mold.

The standard method of cooling is passing a coolant (usually water) through a series of holes drilled through the mold plates and connected by hoses to form a continuous pathway. The coolant absorbs heat from the mold (which has absorbed heat from the hot plastic) and keeps the mold at a proper temperature to solidify the plastic at the most efficient rate.

To ease maintenance and venting, cavities and cores are divided into pieces, called *inserts*, and sub-assemblies, also called *inserts*, *blocks*, or *chase blocks*. By substituting interchangeable inserts, one mold may make several variations of the same part.

More complex parts are formed using more complex molds. These may have sections called slides, that move into a cavity perpendicular to the draw direction, to form overhanging part features. When the mold is opened, the slides are pulled away from the plastic part by using stationary "angle pins" on the stationary mold half. These pins enter a slot in the slides and cause the slides to move backward when the moving half of the mold opens. The part is then ejected and the mold closes. The closing action of the mold causes the slides to move forward along the angle pins.

Some molds allow previously molded parts to be reinserted to allow a new plastic layer to form around the first part. This is often referred to as overmolding. This system can allow for production of one-piece tires and wheels.

Two-shot injection molded keycaps from a computer keyboard.

Two-shot or multi-shot molds are designed to "overmold" within a single molding cycle and must be processed on specialised injection molding machines with two or more injection units. This process is actually an injection molding process performed twice and therefore has a much smaller margin of error. In the first step, the base colour material is molded into a basic shape, which contains spaces for the second shot. Then the second material, a different colour, is injection-molded into those spaces. Pushbuttons and keys, for instance, made by this process have markings that cannot wear off, and remain legible with heavy use.

A mold can produce several copies of the same parts in a single "shot". The number of "impressions" in the mold of that part is often incorrectly referred to as cavitation. A tool with one impression will often be called a single impression (cavity) mold. A mold with 2 or more cavities of the same parts will likely be referred to as multiple impression (cavity) mold. Some extremely high production volume molds (like those for bottle caps) can have over 128 cavities.

In some cases, multiple cavity tooling will mold a series of different parts in the same tool. Some toolmakers call these molds family molds as all the parts are related. Some examples include plastic model kits.

Mold Storage

Manufacturers go to great lengths to protect custom molds due to their high average costs. The perfect temperature and humidity level is maintained to ensure the longest possible lifespan for each custom mold. Custom molds, such as those used for rubber injection molding, are stored in temperature and humidity controlled environments to prevent warping.

Tool Materials

Beryllium-copper insert (yellow) on injection molding mold for ABS resin.

Tool steel is often used. Mild steel, aluminium, nickel or epoxy are suitable only for prototype or very short production runs. Modern hard aluminium (7075 and 2024 alloys) with proper mold design, can easily make molds capable of 100,000 or more part life with proper mold maintenance.

Machining

Molds are built through two main methods: standard machining and EDM. Standard machining, in its conventional form, has historically been the method of building injection molds. With technological developments, CNC machining became the predominant means of making more complex molds with more accurate mold details in less time than traditional methods.

The electrical discharge machining (EDM) or spark erosion process has become widely used in mold making. As well as allowing the formation of shapes that are difficult to machine, the process allows pre-hardened molds to be shaped so that no heat treatment is required. Changes to a hardened mold by conventional drilling and milling normally require annealing to soften the mold, followed by heat treatment to harden it again. EDM is a simple process in which a shaped electrode, usually made of copper or graphite, is very slowly lowered onto the mold surface (over a period of many hours), which is immersed in paraffin oil (kerosene). A voltage applied between tool and mold causes spark erosion of the mold surface in the inverse shape of the electrode.

Cost

The number of cavities incorporated into a mold will directly correlate in molding costs. Fewer cavities require far less tooling work, so limiting the number of cavities in-turn will result in lower initial manufacturing costs to build an injection mold.

As the number of cavities play a vital role in molding costs, so does the complexity of the part's design. Complexity can be incorporated into many factors such as surface finishing, tolerance requirements, internal or external threads, fine detailing or the number of undercuts that may be incorporated.

Further details, such as undercuts or any feature causing additional tooling, will increase the mold cost. Surface finish of the core and cavity of molds will further influence the cost.

Rubber injection molding process produces a high yield of durable products, making it the most efficient and cost-effective method of molding. Consistent vulcanisation processes involving precise temperature control significantly reduces all waste material.

Injection Process

With injection molding, granular plastic is fed by a forced ram from a hopper into a heated barrel. As the granules are slowly moved forward by a screw-type plunger, the plastic is forced into a heated chamber, where it is melted. As the plunger advances, the melted plastic is forced through a nozzle that rests against the mold, allowing it to enter the mold cavity through a gate and runner system. The mold remains cold so the plastic solidifies almost as soon as the mold is filled.

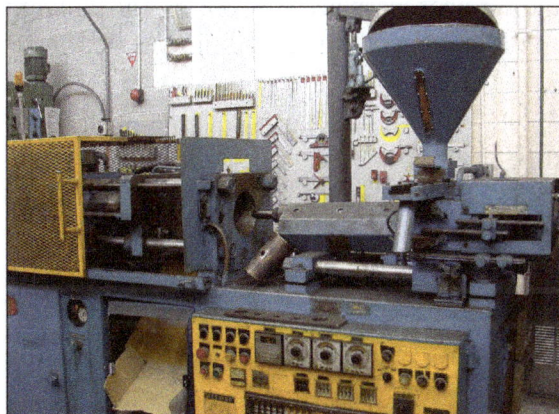

Small injection molder showing hopper, nozzle and die area.

Injection Molding cycle

The sequence of events during the injection mold of a plastic part is called the injection molding cycle. The cycle begins when the mold closes, followed by the injection of the polymer into the mold cavity. Once the cavity is filled, a holding pressure is maintained to compensate for material shrinkage. In the next step, the screw turns, feeding the next shot to the front screw. This causes the screw to retract as the next shot is prepared. Once the part is sufficiently cool, the mold opens and the part is ejected.

Scientific versus Traditional Molding

Traditionally, the injection portion of the molding process was done at one constant pressure to fill and pack the cavity. This method, however, allowed for a large variation in dimensions from cycle-to-cycle. More commonly used now is scientific or decoupled molding, a method pioneered by RJG Inc. In this the injection of the plastic is "decoupled" into stages to allow better control of part dimensions and more cycle-to-cycle (commonly called shot-to-shot in the industry) consistency. First the cavity is filled to approximately 98% full using velocity (speed) control. Although the pressure should be sufficient to allow for the desired speed, pressure limitations during this stage are undesirable. Once the cavity is 98% full, the machine switches from velocity control to pressure control, where the cavity is "packed out" at a constant pressure, where sufficient velocity to reach desired pressures is required. This allows part dimensions to be controlled to within thousandths of an inch or better.

Different Types of Injection Molding Processes

Sandwich-molded toothbrush handle.

Although most injection molding processes are covered by the conventional process description, there are several important molding variations including, but not limited to:

- Die casting.
- Metal injection molding.
- Thin-wall injection molding.
- Injection molding of liquid silicone rubber.
- Reaction injection molding.

Process Troubleshooting

Like all industrial processes, injection molding can produce flawed parts. In the field of injection molding, troubleshooting is often performed by examining defective parts for specific defects and addressing these defects with the design of the mold or the characteristics of the process itself. Trials are often performed before full production runs in an effort to predict defects and determine the appropriate specifications to use in the injection process.

When filling a new or unfamiliar mold for the first time, where shot size for that mold is unknown, a technician/tool setter may perform a trial run before a full production run. They start with a small shot weight and fills gradually until the mold is 95 to 99% full. Once this is achieved, a small amount of holding pressure will be applied and holding time increased until gate freeze off (solidification time) has occurred. Gate freeze off time can be determined by increasing the hold time, and then weighing the part. When the weight of the part does not change, it is then known that the gate has frozen and no more material is injected into the part. Gate solidification time is important, as this determines cycle time and the quality and consistency of the product, which itself is an important issue in the economics of the production process. Holding pressure is increased until the parts are free of sinks and part weight has been achieved.

Molding Defects

Injection molding is a complex technology with possible production problems. They can be caused either by defects in the molds, or more often by the molding process itself.

Molding defects	Alternative name	Descriptions	Causes
Blister	Blistering	Raised or layered zone on surface of the part.	Tool or material is too hot, often caused by a lack of cooling around the tool or a faulty heater.
Burn marks	Air burn/gas burn/dieseling/ gas marks/Blow marks	Black or brown burnt areas on the part located at furthest points from gate or where air is trapped.	Tool lacks venting, injection speed is too high.
Color streaks (US)	Colour streaks (UK)	Localised change of colour	Masterbatch isn't mixing properly, or the material has run out and it's starting to come through as natural only. Previous coloured material "dragging" in nozzle or check valve.
Contamination	Unwanted or foreign material	Different colour matter seen in product, weakening the product.	Poor material introduced by bad recycling or regrind policy; may include floor sweepings, dust and debris.
Delamination		Thin mica like layers formed in part wall.	Contamination of the material e.g. PP mixed with ABS, very dangerous if the part is being used for a safety critical application as the material has very little strength when delaminated as the materials cannot bond.
Flash		Excess material in thin layer exceeding normal part geometry.	Mold is over packed or parting line on the tool is damaged, too much injection speed/material injected, clamping force too low. Can also be caused by dirt and contaminants around tooling surfaces.
Embedded contaminates	Embedded particulates	Foreign particle (burnt material or other) embedded in the part.	Particles on the tool surface, contaminated material or foreign debris in the barrel, or too much shear heat burning the material prior to injection.
Flow marks	Flow lines	Directionally "off tone" wavy lines or patterns.	Injection speeds too slow (the plastic has cooled down too much during injection, injection speeds should be set as fast as is appropriate for the process and material used).
Gate Blush	Halo or Blush Marks	Circular pattern around gate, normally only an issue on hot runner molds.	Injection speed is too fast, gate/sprue/runner size is too small, or the melt/mold temp is too low.

Jetting		Part deformed by turbulent flow of material.	Poor tool design, gate position or runner. Injection speed set too high. Poor design of gates which cause too little die swell and result jetting.
Knit lines	Weld lines	Small lines on the backside of core pins or windows in parts that look like just lines.	Caused by the melt-front flowing around an object standing proud in a plastic part as well as at the end of fill where the melt-front comes together again. Can be minimised or eliminated with a mold-flow study when the mold is in design phase. Once the mold is made and the gate is placed, one can minimise this flaw only by changing the melt and the mold temperature.
Polymer degradation		Polymer breakdown from hydrolysis, oxidation etc.	Excess water in the granules, excessive temperatures in barrel, excessive screw speeds causing high shear heat, material being allowed to sit in the barrel for too long, too much regrind being used.
Sink marks	sinks	Localised depression (In thicker zones).	Holding time/pressure too low, cooling time too short, with sprueless hot runners this can also be caused by the gate temperature being set too high. Excessive material or walls too thick.
Short shot	Short fill or short mold	Partial part	Lack of material, injection speed or pressure too low, mold too cold, lack of gas vents.
Splay marks	Splash mark or silver streaks	Usually appears as silver streaks along the flow pattern, however depending on the type and colour of material it may represent as small bubbles caused by trapped moisture.	Moisture in the material, usually when hygroscopic resins are dried improperly. Trapping of gas in "rib" areas due to excessive injection velocity in these areas. Material too hot, or is being sheared too much.
Stringiness	Stringing or long-gate	String like remnant from previous shot transfer in new shot.	Nozzle temperature too high. Gate hasn't frozen off, no decompression of the screw, no sprue break, poor placement of the heater bands inside the tool.
Voids		Empty space within part (air pocket is commonly used).	Lack of holding pressure (holding pressure is used to pack out the part during the holding time). Filling too fast, not allowing the edges of the part to set up. Also mold may be out of registration (when the two halves don't centre properly and part walls are not the same thickness). The provided information is the common understanding, Correction: The Lack of pack (not holding) pressure (pack pressure is used to pack out even though is the part during the holding time). Filling too fast does not cause this condition, as a void is a sink that did not have a place to happen. In other words, as the part shrinks the resin separated from itself as there was not sufficient resin in the cavity. The void could happen at any area or the part is not limited by the thickness but by the resin flow and thermal conductivity, but it is more likely to happen at thicker areas like ribs or bosses. Additional root causes for voids are unmelt on the melt pool.

Weld line	Knit line / Meld line / Transfer line	Discoloured line where two flow fronts meet.	Mold or material temperatures set too low (the material is cold when they meet, so they don't bond). Time for transition between injection and transfer (to packing and holding) is too early.
Warping	Twisting	Distorted part.	Cooling is too short, material is too hot, lack of cooling around the tool, incorrect water temperatures (the parts bow inwards towards the hot side of the tool) Uneven shrinking between areas of the part.
Cracks	Crazing	Improper fusion of two fluid flows, a state before weld line.	Threadline gap in between part due to improper gate location in complex design parts including excess of holes (multipoint gates to be provided), process optimization, proper air venting.

Methods such as industrial CT scanning can help with finding these defects externally as well as internally.

Tolerances

Molding tolerance is a specified allowance on the deviation in parameters such as dimensions, weights, shapes, or angles, etc. To maximise control in setting tolerances there is usually a minimum and maximum limit on thickness, based on the process used. Injection molding typically is capable of tolerances equivalent to an IT Grade of about 9–14. The possible tolerance of a thermoplastic or a thermoset is ±0.008 to ±0.002 inches. In specialised applications tolerances as low as ±5 µm on both diameters and linear features are achieved in mass production. Surface finishes of 0.0500 to 0.1000 µm or better can be obtained. Rough or pebbled surfaces are also possible.

Molding Type	Typical [mm]	Possible [mm]
Thermoplastic	±0.500	±0.200
Thermoset	±0.500	±0.200

Power Requirements

The power required for this process of injection molding depends on many things and varies between materials used. *Manufacturing Processes Reference Guide* states that the power requirements depend on "a material's specific gravity, melting point, thermal conductivity, part size, and molding rate." Below is a table from page 243 of the same reference as previously mentioned that best illustrates the characteristics relevant to the power required for the most commonly used materials.

Material	Specific gravity	Melting point (°F)	Melting point (°C)
Epoxy	1.12 to 1.24	248	120
Phenolic	1.34 to 1.95	248	120
Nylon	1.01 to 1.15	381 to 509	194 to 265
Polyethylene	0.91 to 0.965	230 to 243	110 to 117
Polystyrene	1.04 to 1.07	338	170

Robotic Molding

Automation means that the smaller size of parts permits a mobile inspection system to examine multiple parts more quickly. In addition to mounting inspection systems on automatic devices, multiple-axis robots can remove parts from the mold and position them for further processes.

Specific instances include removing of parts from the mold immediately after the parts are created, as well as applying machine vision systems. A robot grips the part after the ejector pins have been extended to free the part from the mold. It then moves them into either a holding location or directly onto an inspection system. The choice depends upon the type of product, as well as the general layout of the manufacturing equipment. Vision systems mounted on robots have greatly enhanced quality control for insert molded parts. A mobile robot can more precisely determine the placement accuracy of the metal component, and inspect faster than a human can.

Reaction Injection Molding

Reaction injection molding (RIM) is similar to injection molding except thermosetting polymers are used, which requires a curing reaction to occur within the mold.

Common items made via RIM include automotive bumpers, air spoilers, and fenders.

Process

The two parts of the polymer are mixed together, usually by injecting them under high pressure into an impinging mixer. Then the mixture is injected under lower pressure into a mold. The mixture is allowed to sit in the mold long enough for it to expand and cure.

If reinforcing agents are added to the mixture then the process is known as reinforced reaction injection molding (RRIM). Common reinforcing agents include glass fibers and mica. This process is usually used to produce rigid foam automotive panels.

A subset of RIM is structural reaction injection molding (SRIM), which uses fiber meshes for the reinforcing agent. The fiber mesh is first arranged in the mold and then the polymer mixture is injection molded over it.

The most common RIM processable material is polyurethane (known generally as PU-RIM), but others include polyureas, polyisocyanurates, polyesters, polyphenols, polyepoxides, and nylon 6. For polyurethane one component of the mixture is polyisocyanate and the other component is a blend of polyol, surfactant, catalyst, and blowing agent.

Advantages and Disadvantages

Reaction injection molding can produce strong, flexible, lightweight parts which can easily be painted. It also has the advantage of quick cycle times compared to typical vacuum cast materials. The bi-component mixture injected into the mold has a much lower viscosity than molten thermoplastic polymers, therefore large, light-weight, and thin-walled items can be successfully RIM processed. This thinner mixture also requires less clamping forces, which leads to smaller equipment and ultimately lower capital expenditures. Another advantage of RIM processed foam is that a high-density skin is formed with a low-density core.

The disadvantages are slow cycle times, compared to injection molding, and expensive raw materials.

Tooling

Machined steel or aluminum; cast aluminum; silicone rubber; epoxy resin; nickel. The machines can be large or small depending on the size of part required.

Rotational Molding

A three-motor powered (tri-power) rotational-molding or spin-casting machine.

Rotational Molding (BrE molding) involves a heated hollow mold which is filled with a charge or shot weight of material. It is then slowly rotated (usually around two perpendicular axes), causing the softened material to disperse and stick to the walls of the mold. In order to maintain even thickness throughout the part, the mold continues to rotate at all times during the heating phase and to avoid sagging or deformation also during the cooling phase. The process was applied to plastics in the 1950s but in the early years was little used because it was a slow process restricted to a small number of plastics. Over time, improvements in process control and developments with plastic powders have resulted in a significant increase in usage.

Rotocasting (also known as rotacasting), by comparison, uses self-curing resins in an unheated mold, but shares slow rotational speeds in common with rotational molding. Spin casting should not be confused with either, utilizing self-curing resins or white metal in a high-speed centrifugal casting machine.

Equipment and Tooling

Rotational molding machines are made in a wide range of sizes. They normally consist of molds, an oven, a cooling chamber, and mold spindles. The spindles are mounted on a rotating axis, which provides a uniform coating of the plastic inside each mold.

Molds (or tooling) are either fabricated from welded sheet steel or cast. The fabrication method is often driven by part size and complexity; most intricate parts are likely made out of cast tooling. Molds are typically manufactured from stainless steel or aluminum. Aluminum molds are usually much thicker than an equivalent steel mold, as it is a softer metal. This thickness does not affect cycle times significantly since aluminum's thermal conductivity is many times greater than steel.

Due to the need to develop a model prior to casting, cast molds tend to have additional costs associated with the manufacturing of the tooling, whereas fabricated steel or aluminum molds, particularly when used for less complex parts, are less expensive. However, some molds contain both aluminum and steel. This allows for variable thicknesses in the walls of the product. While this process is not as precise as injection molding, it does provide the designer with more options. The aluminum addition to the steel provides more heat capacity, causing the melt-flow to stay in a fluid state for a longer period.

Standard Setup and Equipment for Rotational Molding

Normally all rotation molding systems have a number of parts including molds, oven, cooling chamber and mold spindles. The molds are used to create the part, and are typically made of aluminium. The quality and finish of the product is directly related to the quality of the mold being used. The oven is used to heat the part while also rotating the part to form the part desired. The cooling chamber is where the part is placed until it cools, and the spindles are mounted to rotate and provide a uniform coat of plastic inside each mold.

Rotational Molding Machines

Rock and Roll Machine

This is a specialized machine designed mainly to produce long narrow parts. Some are of the clamshell type, thus one arm, but there are also shuttle-type Rock & Roll machines, with two arms. Each arm rotates or rolls the mold 360 degrees in one direction and at the same time tips and rocks the mold 45 degrees above or below horizontal in the other direction. Newer machines use forced hot air to heat the mold. These machines are best for large parts that have a large length-to-width ratio. Because of the smaller heating chambers, there is a saving in heating costs compared to bi-axial machines.

A Rock and Roll rotational molding machine built in 2009.

Clamshell Machine

This is a single arm rotational molding machine. The arm is usually supported by other arms on both ends. The clamshell machine heats and cools the mold in the same chamber. It takes up less space than equivalent shuttle and swing arm rotational molders. It is low in cost compared to the size of products made. It is available in smaller scales for schools interested in prototyping and for high quality models. More than one mold can be attached to the single arm.

Vertical or Up and Over Rotational Machine

The loading and unloading area is at the front of the machine between the heating and cooling areas. These machines vary in size between small to medium compared to other rotational machines. Vertical rotational molding machines are energy efficient due to their compact heating and cooling chambers. These machines have the same (or similar) capabilities as the horizontal carousel multi-arm machines, but take up much less space.

Shuttle Machine

Most shuttle machines have two arms that move the molds back and forth between the heating chamber and cooling station. The arms are independent of each other and they turn the molds bi-axially. In some cases, the shuttle machine has only one arm. This machine moves the mold in a linear direction in and out of heating and cooling chambers. It is low in cost for the size of product produced and the footprint is kept to a minimum compared to other types of machines. It is also available in smaller scale for schools and prototyping.

Swing Arm Machine

The swing-arm machine can have up to four arms, with a bi-axial movement. Each arm is independent from each other as it is not necessary to operate all arms at the same time. Each arm is mounted on a corner of the oven and it swings in and out of the oven. On some swing-arm machines, a pair of arms is mounted on the same corner, thus a four-arm machine has two pivot points. These machines are very useful for companies that have long cooling cycles or require a lot of time to demold parts, compared to the cook time. It is a lot easier to schedule maintenance work or try to run a new mold without interrupting production on the other arms of the machine.

Carousel Machine

A Carousel machine with four independent arms.

This is one of the most common bi-axial machines in the industry. It can have up to 4 arms and six stations and it comes in a wide range of sizes. The machine comes in two different models, fixed and independent. A fixed-arm carousel consists of 3 fixed arms that must move together. One arm will be in the heating chamber while the other is in the cooling chamber and the other in the loading/reloading area. The fixed-arm carousel works well when working with identical cycle times on each arm. The independent-arm carousel machine is available with 3 or 4 arms that can move separately from the other. This allows for different size molds, with different cycle times and thickness needs.

Production Process

The rotational molding process is a high-temperature, low-pressure plastic-forming process that uses heat and biaxial rotation (i.e., angular rotation on two axes) to produce hollow, one-piece parts. Critics of the process point to its long cycle times—only one or two cycles an hour can typically occur, as opposed to other processes such as injection molding, where parts can be made in a few seconds. The process does have distinct advantages. Manufacturing large, hollow parts such as oil tanks is much easier by rotational molding than any other method. Rotational molds are significantly cheaper than other types of mold. Very little material is wasted using this process, and excess material can often be re-used, making it a very economically and environmentally viable manufacturing process.

Unloading a molded polyethylene tank in a Shuttle machine.

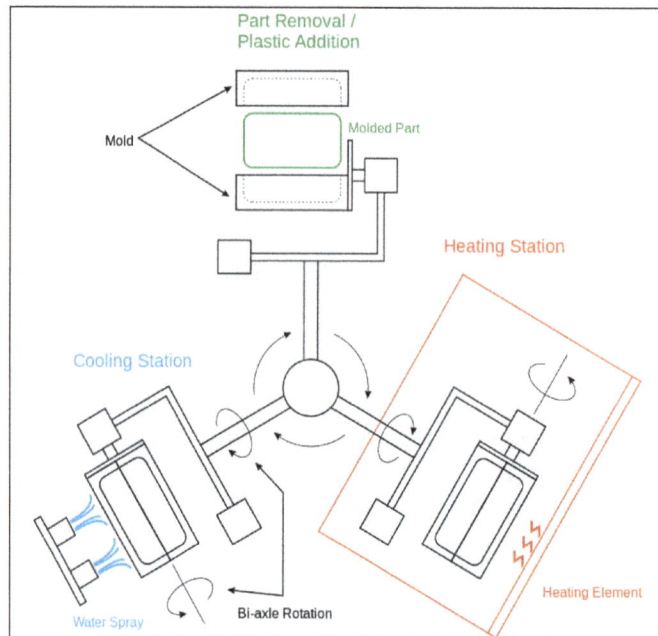

Rotational Molding Process.

The rotational molding process consists of four distinct phases:

- Loading a measured quantity of polymer (usually in powder form) into the mold.

- Heating the mold in an oven while it rotates, until all the polymer has melted and adhered to the mold wall. The hollow part should be rotated through two or more axes, rotating at different speeds, in order to avoid the accumulation of polymer powder. The length of time the mold spends in the oven is critical: too long and the polymer will degrade, reducing impact strength. If the mold spends too little time in the oven, the polymer melt may be incomplete. The polymer grains will not have time to fully melt and coalesce on the mold wall, resulting in large bubbles in the polymer. This has an adverse effect on the mechanical properties of the finished product.

- Cooling the mold, usually by fan. This stage of the cycle can be quite lengthy. The polymer must be cooled so that it solidifies and can be handled safely by the operator. This typically takes tens of minutes. The part will shrink on cooling, coming away from the mold, and facilitating easy removal of the part. The cooling rate must be kept within a certain range. Very rapid cooling (for example, water spray) would result in cooling and shrinking at an uncontrolled rate, producing a warped part.

- Removal of the part.

Recent Improvements

Until recently, the process largely relied on both trial and error and the experience of the operator to determine when the part should be removed from the oven and when it was cool enough to be removed from the mold. Technology has improved in recent years, allowing the air temperature in the mold to be monitored, removing much of the guesswork from the process.

Much of the current research is into reducing the cycle time, as well as improving part quality. The most promising area is in mold pressurization. It is well known that applying a small amount of pressure internally to the mold at the correct point in the heating phase accelerates coalescence of the polymer particles during the melting, producing a part with fewer bubbles in less time than at atmospheric pressure. This pressure delays the separation of the part from the mold wall due to shrinkage during the cooling phase, aiding cooling of the part. The main drawback to this is the danger to the operator of explosion of a pressurized part. This has prevented adoption of mold pressurization on a large scale by rotomolding manufacturers.

Mold Release Agents

A good mold release agent (MRA) will allow the material to be removed quickly and effectively. Mold releases can reduce cycle times, defects, and browning of finished product. There are a number of mold release types available; they can be categorized as follows:

- Sacrificial coatings: The coating of MRA has to be applied each time because most of the MRA comes off on the molded part when it releases from the tool. Silicones are typical MRA compounds in this category.

- Semi-permanent coatings: The coating, if applied correctly, will last for a number of releases before requiring to be re-applied or touched up. This type of coating is most prevalent in today's rotational molding industry. The active chemistry involved in these coatings is typically a polysiloxane.

- Permanent coatings: Most often some form of PTFE coating, which is applied to the mold. Permanent coatings avoid the need for operator application, but may become damaged by misuse.

Materials

More than 80% of all the material used is from the polyethylene family: cross-linked polyethylene (PEX), low-density polyethylene (LDPE), linear low-density polyethylene (LLDPE), high-density polyethylene (HDPE), and regrind. Other compounds are PVC plastisols, nylons, and polypropylene.

Order of materials most commonly used by industry:

- Polyethylene,
- Polypropylene,
- Polyvinyl chloride,
- Nylon,
- Polycarbonate.

These materials are also occasionally used (not in order of most used):

- Aluminum,
- Acrylonitrile butadiene styrene (ABS),
- Acetal,
- Acrylic,
- Epoxy,
- Fluorocarbons,
- Ionomer,
- Polybutylene,
- Polyester,
- Polystyrene,
- Polyurethane,
- Silicone,
- Various foods (especially chocolate).

Natural Materials

Recently it has become possible to use natural materials in the molding process. Through the use of real sands and stone chip, sandstone composite can be created which is 80% natural non-processed material.

Rotational molding of plaster is used to produce hollow statuettes.

Chocolate is rotationally molded to form hollow treats.

Products

Designers can select the best material for their application, including materials that meet U.S. Food and Drug Administration (FDA) requirements. Additives for weather resistance, flame retardation, or static elimination can be incorporated. Inserts, graphics, threads, handles, minor undercuts, flat surfaces without draft angles, or fine surface detail can be part of the design. Designs can also be multi-wall, either hollow or foam filled.

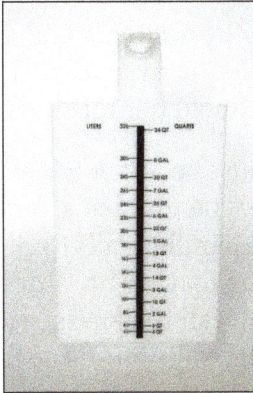

Mold in graphic.

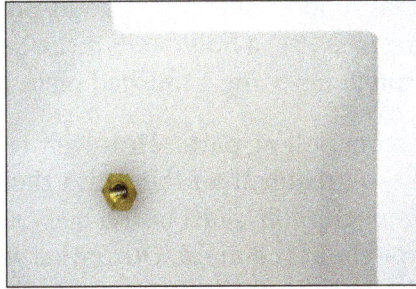

A blind brass threaded hex *insert* molded into a liquid storage tank.

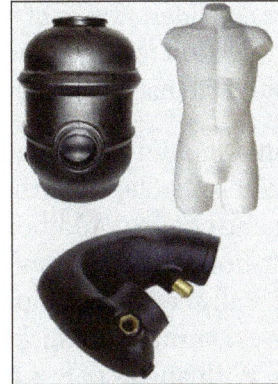

Products that can be manufactured using rotational molding include storage tanks, furniture, road signs and bollards, planters, pet houses, toys, bins and refuse containers, doll parts, road cones, footballs, helmets, canoes, rowing boats, tornado shelters, kayak hulls, underground cellars for vine and vegetables storage and playground slides. The process is also used to make highly specialised products, including UN-approved containers for the transportation of nuclear fissile materials, anti-piracy ship protectors, seals for inflatable oxygen masks and lightweight components for the aerospace industry.

Rotational Molded Flamingo.

Edon roto molded rowing boat.

Design Considerations

Designing for Rotational Molding

Another consideration is in the draft angles. These are required to remove the piece from the mold. On the outside walls, a draft angle of 1° may work (assuming no rough surface or holes). On inside walls, such as the inside of a boat hull, a draft angle of 5° may be required. This is due to shrinkage and possible part warping.

Another consideration is of structural support ribs. While solid ribs may be desirable and achievable in injection molding and other processes, a hollow rib is the best solution in rotational molding. A solid rib may be achieved through inserting a finished piece in the mold but this adds cost.

Rotational molding excels at producing hollow parts. However, care must be taken when this is done. When the depth of the recess is greater than the width there may be problems with even heating and cooling. Additionally, enough room must be left between the parallel walls to allow for the melt-flow to move properly throughout the mold. Otherwise webbing may occur. A desirable parallel wall scenario would have a gap at least three times the nominal wall thickness, with five times the nominal wall thickness being optimal. Sharp corners for parallel walls must also be considered. With angles of less than 45° bridging, webbing, and voids may occur.

Material Limitations and Considerations

Another consideration is the melt-flow of materials. Certain materials, such as nylon, will require larger radii than other materials. Additionally, the stiffness of the set material may be a factor. More structural and strengthening measures may be required when a flimsy material is used.

Wall Thickness

One benefit of rotational molding is the ability to experiment, particularly with wall thicknesses. Cost is entirely dependent on wall thickness, with thicker walls being costlier and more time consuming to produce. While the wall can have nearly any thickness, designers must remember that the thicker the wall, the more material and time will be required, increasing costs. In some cases, the plastics may significantly degrade due to extended periods at high temperature. Also, different materials have different thermal conductivity, meaning they require different times in the heating chamber and cooling chamber. Ideally, the part will be tested to use the minimum thickness required for the application. This minimum will then be established as a nominal thickness.

For the designer, while variable thicknesses are possible, a process called stop rotation is required. This process is limited in that only one side of the mold may be thicker than the others. After the mold is rotated and all the surfaces are sufficiently coated with the melt-flow, the rotation stops and the melt-flow is allowed to pool at the bottom of the mold cavity.

Wall thickness is important for corner radii as well. Large outside radii are preferable to small radii. Large inside radii are also preferable to small inside radii. This allows for a more even flow of material and a more even wall thickness. However, an outside corner is generally stronger than an inside corner.

Advantages

Rotational molding offers design advantages over other molding processes. With proper design, parts assembled from several pieces can be molded as one part, eliminating high fabrication costs. The process also has inherent design strengths, such as consistent wall thickness and strong outside corners that are virtually stress free. For additional strength, reinforcing ribs can be designed into the part. Along with being designed into the part, they can be added to the mold.

The ability to add prefinished pieces to the mold alone is a large advantage. Metal threads, internal pipes and structures, and even different colored plastics can all be added to the mold prior to the addition of plastic pellets. However, care must be taken to ensure that minimal shrinkage while cooling will not damage the part. This shrinking allows for mild undercuts and negates the need for ejection mechanisms (in most pieces).

In some cases rotational molding can be used as a feasible alternative to blow molding, this is due to the similarity in product outputs, with products such as plastic bottles and cylindrical containers, this is only effective on a smaller scale as it much more costly to blow mold regarding a small output, and with fewer resulting products rotational molding is much cheaper, due to blow molding relying on economies of scale regarding efficiency.

Another advantage lies in the molds themselves. Since they require less tooling, they can be manufactured and put into production much more quickly than other molding processes. This is especially true for complex parts, which may require large amounts of tooling for other molding processes. Rotational molding is also the process of choice for short runs and rush deliveries. The molds can be swapped quickly or different colors can be used without purging the mold. With other processes, purging may be required to swap colors.

Due to the uniform thicknesses achieved, large stretched sections are nonexistent, which makes large thin panels possible (although warping may occur). Also, there is little flow of plastic (stretching) but rather a placing of the material within the part. These thin walls also limit cost and production time.

Another cost advantage with rotational molding is the minimal amount of material wasted in production. There are no sprues or runners (as in injection molding), no off-cuts , or pinch off scrap (blow molding). What material is wasted, through scrap or failed part testing, can usually be recycled.

Disadvantages

Rotationally molded parts have to follow some restrictions that are different from other plastic processes. As it is a low pressure process, sometimes designers face hard to reach areas in the mold. Good quality powder may help overcome some situations, but usually the designers have to keep in mind that it is not possible to make sharp threads that would be possible with injection molding. Some products based on polyethylene can be put in the mold before filling it with the main material. This can help to avoid holes that otherwise would appear in some areas. This could also be achieved using molds with movable sections.

Another limitation lies in the molds themselves. Unlike other processes where only the product needs to be cooled before being removed, with rotational molding the entire mold must be cooled.

While water cooling processes are possible, there is still a significant down time of the mold. Additionally, this increases both financial and environmental costs. Some plastics will degrade with the long heating cycles or in the process of turning them into a powder to be melted.

The stages of heating and cooling involve transfer of heat first from the hot medium to the polymer material and next from it to the cooling environment. In both cases, the process of heat transfer occurs in an unsteady regime; therefore, its kinetics attracts the greatest interest in considering these steps. In the heating stage, the heat taken from the hot gas is absorbed both by the mold and the polymer material. The rig for rotational molding usually has a relatively small wall thickness and is manufactured from metals with a high thermal conductivity (aluminum, steel). As a rule, the mold transfers much more heat than plastic can absorb; therefore, the mold temperature must vary linearly. The rotational velocity in rotational molding is rather low (4 to 20 rpm). As a result, in the first stages of the heating cycle, the charged material remains as a powder layer at the bottom of the mold. The most convenient way of changing the cycle is by applying PU sheets in hot rolled forms.

Material Requirements

Due to the nature of the process, materials selection must take into account the following:

- Due to high temperatures within the mold the plastic must have a high resistance to permanent change in properties caused by heat (high thermal stability).

- The molten plastic will come into contact with the oxygen inside the mold—this can potentially lead to oxidation of the melted plastic and deterioration of the material's properties. Therefore, the chosen plastic must have a sufficient amount of antioxidant molecules to prevent such degradation in its liquid state.

- Because there is no pressure to push the plastic into the mold, the chosen plastic must be able to flow easily through the cavities of the mold. The part's design must also take into account the flow characteristics of the particular plastic chosen.

Rotational molding is basically used for big storage tanks.

Transfer Molding

Transfer molding (BrE molding) is a manufacturing process where casting material is forced into a mold. Transfer molding is different from compression molding in that the mold is enclosed [Hayward] rather than open to the fill plunger resulting in higher dimensional tolerances and less environmental impact. Compared to injection molding, transfer molding uses higher pressures to uniformly fill the mold cavity. This allows thicker reinforcing fiber matrices to be more completely saturated by resin. Furthermore, unlike injection molding the transfer mold casting material may start the process as a solid. This can reduce equipment costs and time dependency. The transfer process may have a slower fill rate than an equivalent injection molding processes.

Process

The mold interior surfaces may be gel-coated. If desired the mold is first pre-loaded with a reinforcing fiber matrix or preform. Fiber content of a transfer molded composite can be as high as

60% by volume. The fill material may be a preheated solid or a liquid. It is loaded into a chamber known as the pot. A ram or plunger forces material from the pot into the heated mold cavity. If feed-stock is initially solid, the forcing pressure and mold temperature melt it. Standard mold features such as sprue channels, a flow gate and ejector pins may be used. The heated mold ensures that the flow remains liquid for complete filling. Once filled the mold can be cooled at a controlled rate for optimal thermoset curing.

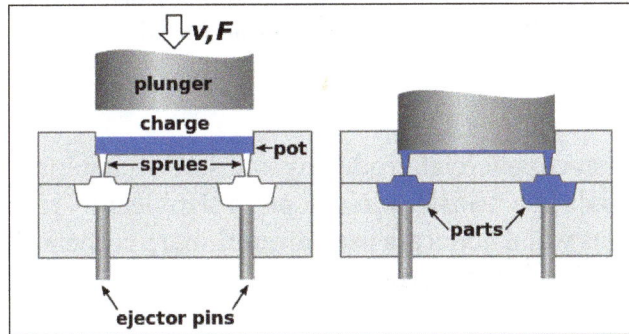

Transfer Molding basic process.

Variations

The industry identifies a variety of processes within the transfer molding category. There are areas of overlap and the distinctions between each method may not be clearly defined.

Resin Transfer Molding

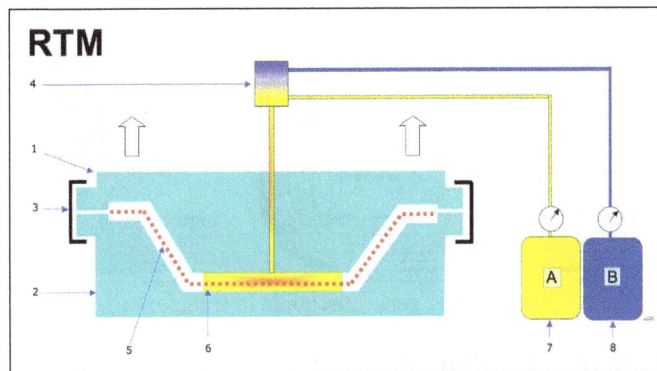

Resin Transfer Molding 1: Cope 2: Drag 3: Clamp 4: Mixing chamber
5: Fiber preform 6: Heated mold 7: Resin 8: Curative.

Resin transfer molding (RTM) uses a liquid thermoset resin to saturate a fiber preform placed in a closed mold. The process is versatile and can fabricate products with embedded objects such as foam cores or other components in addition to the fiber preform.

Vacuum Assisted Resin Transfer Molding

Vacuum assisted transfer molding (VARTM) uses a partial vacuum on one side of a fiber mat to pull the resin in for complete saturation. VARTM uses lower plunger forces which allows molding to be carried out with cheaper equipment. The use of a vacuum may allow the resin to adequately flow and or cure without heating. This temperature independence allows thicker fiber preforms and larger product geometries to be economical. VARTM can produce parts with less porosity than regular transfer molding with a proportional increase in casting strength.

Micro Transfer Molding

Also called transfer micromolding, this is a process that uses a mold to form then transfer structures as small as 30 nm onto thin films and microcircuitry. Unlike normal scale transfer molding, the micro form can and is used with metals as well as non metals.

Defects

Limiting defects is key when commercial producing any sort of material. Transfer molding is no exception. For example, voids in a transfer molded parts significantly reduce strength and modulus. There can also be defects when fibers are used around sharp corners. The resin flow can create resin rich zones on the outside of these corners.

Pressure Distribution

There are several contributing factors to voids in the final product of transfer molding. One is a non uniform pressure distribution among the material being pressed into the mold. In this case the material folds in on itself and generates voids. Another is voids in the resin being forced into the mold before hand. This maybe obvious, but it is a main contributor. Things to be done to limit these molds include pressing the resin in at a high pressure, keeping the fiber distribution uniform, and using a high quality properly degassed base resin.

Sharp Corners

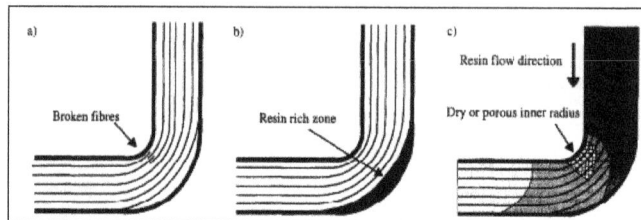
Figure: Sharp corner generates voids in transfer molding.

Sharp corners are the problems with all mold based manufacturing, including casting. Specifically in transfer molding corners can break fibers that have been placed in the mold and can create voids on the inside of corners. The limiting factor in these designs is the inner corner radius. This inner radius limit varies depending on resin and fiber selection, but a rule of thumbs is the radius though be 3 to 5 times the laminate thickness.

Materials

The material most commonly used for transfer molding is a thermoset polymer. This type of polymer is easy to mold and manipulate, but upon curing, hardens into a permanent form. For simple homogeneous transfer molded parts, the part is simply made of this plastic substrate. On the other hand, resin transfer molding allows for a composite material to be made by placing a fiber within the mold and subsequently injecting the thermosetting polymer.

Defects known as voids and dry resin (in the case of resin transfer molding) are possible in transfer molding and often are exacerbated by high viscosity materials. This is because a high viscosity plastic flowing through a thin mold may miss entire vacated areas, leaving air pockets. When air pockets are left in the presence of fiber, this creates a "dry" area, which prevents load from being transferred through the fibers in the dry area.

Materials used for the plastic are often polyurethanes or epoxy resins. Both of these are soft and malleable before curing, becoming much harder after setting. Materials used for fibers vary extensively, although common choices are carbon or Kevlar fibers, as well as organic fibers, such as hemp.

Compression Molding

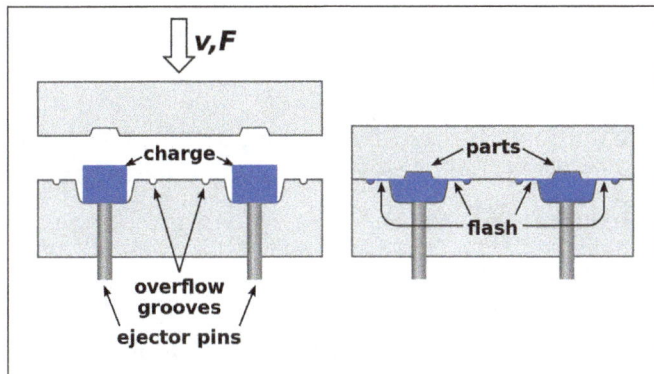

Compression molding - simplified diagram of the process.

Compression molded rubber boots before the flashes are removed.

Compression Molding is a method of molding in which the molding material, generally preheated, is first placed in an open, heated mold cavity. The mold is closed with a top force or plug member, pressure is applied to force the material into contact with all mold areas, while heat and pressure are maintained until the molding material has cured. The process employs thermosetting resins in a partially cured stage, either in the form of granules, putty-like masses, or preforms.

Compression molding is a high-volume, high-pressure method suitable for molding complex, high-strength fiberglass reinforcements. Advanced composite thermoplastics can also be compression molded with unidirectional tapes, woven fabrics, randomly oriented fiber mat or chopped strand. The advantage of compression molding is its ability to mold large, fairly intricate parts. Also, it is one of the lowest cost molding methods compared with other methods such as transfer molding and injection molding; moreover it wastes relatively little material, giving it an advantage when working with expensive compounds.

However, compression molding often provides poor product consistency and difficulty in controlling flashing, and it is not suitable for some types of parts. Fewer knit lines are produced and a smaller amount of fiber-length degradation is noticeable when compared to injection molding. Compression-molding is also suitable for ultra-large basic shape production in sizes beyond the capacity of extrusion techniques. Materials that are typically manufactured through compression molding include: Polyester fiberglass resin systems (SMC/BMC), Torlon, Vespel, Poly(p-phenylene sulfide) (PPS), and many grades of PEEK.

Compression molding is commonly utilized by product development engineers seeking cost effective rubber and silicone parts. Manufacturers of low volume compression molded components include PrintForm, 3D, STYS, and Aero MFG.

Compression molding was first developed to manufacture composite parts for metal replacement applications, compression molding is typically used to make larger flat or moderately curved parts. This method of molding is greatly used in manufacturing automotive parts such as hoods, fenders, scoops, spoilers, as well as smaller more intricate parts. The material to be molded is positioned in the mold cavity and the heated platens are closed by a hydraulic ram. Bulk molding compound (BMC) or sheet molding compound (SMC), are conformed to the mold form by the applied pressure and heated until the curing reaction occurs. SMC feed material usually is cut to conform to the surface area of the mold. The mold is then cooled and the part removed.

Mold and material behavior vary based on material types. New I-PRESS Servo Hydraulic technology provides end users with a greater degree of flexibility with press movement & pressure profiles and control of external devices such as automation to platen zone temperature control. With the latest of electronic servo motor to high pressure hydraulic pump users save considerable energy and the greatest in speed, distance, pressure, dwell time and burping movements.

Molding press.

Materials may be loaded into the mold either in the form of pellets or sheet, or the mold may be loaded from a plasticating extruder. Materials are heated above their melting points, formed and cooled. The more evenly the feed material is distributed over the mold surface, the less flow orientation occurs during the compression stage.

Compression molding is also widely used to produce sandwich structures that incorporate a core material such as a honeycomb or polymer foam.

Thermoplastic matrices are commonplace in mass production industries. One significant example are automotive applications where the leading technologies are long fibre reinforced thermoplastics (LFT) and glass fiber mat reinforced thermoplastics (GMT).

In compression molding there are six important considerations that an engineer should bear in mind:

- Determining the proper amount of material.

- Determining the minimum amount of energy required to heat the material.

- Determining the minimum time required to heat the material.

- Determining the appropriate heating technique.

- Predicting the required force, to ensure that shot attains the proper shape.

- Designing the mold for rapid cooling after the material has been compressed into the mold.

Compression molding is a forming process in which a plastic material is placed directly into a heated metal mold then is softened by the heat and therefore forced to conform to the shape of the mold, as the mold closes. Once molding is completed excess Flash may be removed. Typically, compression molding machines open along a vertical axis.

Process Characteristics

The use of thermoset plastic compounds characterizes this molding process from many of the other molding processes. These thermosets can be in either preform or granule shapes. Unlike some of the other processes we find that the materials are usually preheated and measured before molding. This helps to reduce excess flash. Inserts, usually metallic, can also be molded with the plastic. As a side note, remember not to allow any undercuts on the shape, it will make ejection especially difficult.

Process Schematic

Compression molding is one of the oldest manufacturing technique for rubber molding. The process parameters includes molding time, temperature, and pressure. Usually, a 300-400 ton clamp pressure is used. The typical mold is shaped like a clam shell. The molding press looked a lot like a ladle filled vertical press used for casting aluminum. The bottom of the mold was always the cavity. Compression molding used preforms made by an extruder/wink cutter or a roller die/die cutter. Wink meaning that 2 blades meet on center to cut the extrudate to length. For example; Molding water bottles used die cut sheets from a roller die. The sheet was 3 inch by 6 inch. The first sheet was placed- one below a core and one sheet of equal size above the core, and then the top of the mold lowered by hand or by hoist to near shut. The mold was then pushed into the press. The start button hydraulically closed the vertical press to full pressure. The mold temperature was about 350 degrees. The platens of the presses were steam heated.

When the cycle ended (about 3.5-4.0 Minutes) the press would open and the mold would be pulled out toward the operator. The operator would try to open the clam shell mold top, and then lean the top of the mold back against the press. Exposed is the bottle with the core still inside. While the bottle was still hot the operator would insert prongs like reverse pliers in between the bottle rubber and the steel core. The operator would then stretch the bottle at the neck over the core to free the bottle. In preparation of compression molding baby nipples and golf ball centers the pre-forms were extruded. The baby nipple was a kidney shape about 2 inches tall and 1/2 inch wide in the middle. The golf ball center preform had a 1 x 1 inch round slug. Both slugs were designed to stand up in the mold cavity. During the cycle the operator would load the jig with slugs. When the mold is opened, the lower platen would lower and the mold would be hydraulically pushed ou to the operator. Therein, the heat sheet (all molded parts from that cycle were joined together by a parting line rind (flash)) were placed in a transfer cart.

The next cycle began by the jig being put over the mold. The slide tray was pulled and the preforms were released into the cavity of the mold. The start button moved the lower platen back into the press and the cure cycle bagan again. Therein the first cycle was complete. Each operator ran an average of 4 presses. Loading an unload was done during the cycle. The heat sheets removed from the mold were then transported to a die station. The die out station would remove the rind leaving the finished parts. With the evolution of compression molding, next was injection transfer. Basically the extruder was made part of the molding cycle. The rubber was injected into an upper heated platen station, and then pressure was applied to transfer molten rubber to the clamped mold. The design of injection transfer and improved molds were more so plastic injection molding except the platens and molds of injection transfer are heated. In contrast plastic Injections molding shoots a hot plastic into a cold mold.

METAL SPINNING

Metal spinning, also known as spin forming or spinning or metal turning most commonly, is a metalworking process by which a disc or tube of metal is rotated at high speed and formed into an axially symmetric part. Spinning can be performed by hand or by a CNC lathe.

Metal spinning does not involve removal of material, as in conventional wood or metal turning, but forming (molding) of sheet material over an existing shape.

Metal spinning ranges from an artisan's specialty to the most advantageous way to form round metal parts for commercial applications. Artisans use the process to produce architectural detail, specialty lighting, decorative household goods and urns. Commercial applications include rocket nose cones, cookware, gas cylinders, brass instrument bells, and public waste receptacles. Virtually any ductile metal may be formed, from aluminum or stainless steel, to high-strength, high-temperature alloys including INX, Inconel, Grade 50 / Corten, and Hastelloy. The diameter and depth of formed parts are limited only by the size of the equipment available.

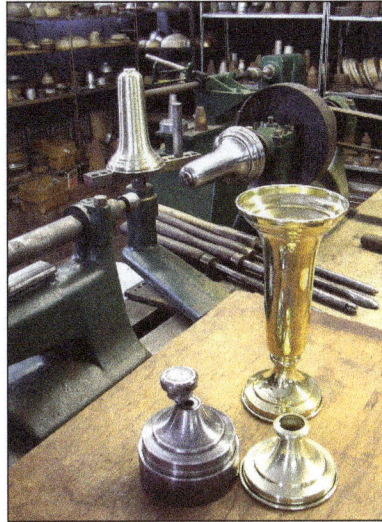

A brass vase spun by hand. Mounted to the lathe spindle is the mandrel for the body of the vase; a shell sits on the "T" rest. The foreground shows the mandrel for the base. Behind the finished vase are the spinning tools used to shape the metal.

Process

The spinning process is fairly simple. A formed block is mounted in the drive section of a lathe. A pre-sized metal disk is then clamped against the block by a pressure pad, which is attached to the tailstock. The block and workpiece are then rotated together at high speeds. A localized force is then applied to the workpiece to cause it to flow over the block. The force is usually applied via various levered tools. Simple workpieces are just removed from the block, but more complex shapes may require a multi-piece block. Extremely complex shapes can be spun over ice forms, which then melt away after spinning. Because the final diameter of the workpiece is always less than the starting diameter, the workpiece must thicken, elongate radially, or buckle circumferentially.

A more involved process, known as *reducing* or *necking*, allows a spun workpiece to include reentrant geometries. If surface finish and form are not critical, then the workpiece is "spun on air"; no mandrel is used. If the finish or form are critical then an eccentrically mounted mandrel is used.

"Hot spinning" involves spinning a piece of metal on a lathe while high heat from a torch is applied to the workpiece. Once heated, the metal is then shaped as the tool on the lathe presses against the heated surface forcing it to distort as it spins. Parts can then be shaped or necked down to a smaller diameter with little force exerted, providing a seamless shoulder.

Tools

The basic hand metal spinning tool is called a *spoon*, though many other tools (be they commercially produced, ad hoc, or improvised) can be used to effect varied results. Spinning tools can be made of hardened steel for use with aluminum, or from solid brass for spinning stainless steel or mild steel.

Some metal spinning tools are allowed to spin on bearings during the forming process. This reduces friction and heating of the tool, extending tool life and improving surface finish. Rotating tools may also be coated with thin film of ceramic to prolong tool life. Rotating tools are commonly used during CNC metal spinning operations.

Commercially, rollers mounted on the end of levers are generally used to form the material down to the mandrel in both hand spinning and CNC metal spinning. Rollers vary in diameter and thickness depending the intended use. The wider the roller the smoother the surface of the spinning; the thinner rollers can be used to form smaller radii.

Cutting of the metal is done by hand held cutters, often foot long hollow bars with tool steel shaped/sharpened files attached. In CNC applications, carbide or tool steel cut-off tools are used.

The mandrel does not incur excessive forces, as found in other metalworking processes, so it can be made from wood, plastic, or ice. For hard materials or high volume use, the mandrel is usually made of metal.

Advantages and Disadvantages

Several operations can be performed in one set-up. Work pieces may have re-entrant profiles and the profile in relation to the center line virtually unrestricted.

Forming parameters and part geometry can be altered quickly, at less cost than other metal forming techniques. Tooling and production costs are also comparatively low. Spin forming, often done by hand, is easily automated and an effective production method for prototypes as well as high quantity production runs.

Other methods of forming round metal parts include hydroforming, stamping, forging and casting. These other methods generally have a higher fixed cost, but a lower variable cost than metal spinning. As machinery for commercial applications has improved, parts are being spun with thicker materials in excess of 1in (25mm) thick steel. Conventional spinning also wastes a considerably smaller amount of material than other methods.

Objects can be built using one piece of material to produce parts without seams. Without seams, a part can withstand higher internal or external pressure exerted on it. For example: scuba tanks and CO_2 cartridges.

One disadvantage of metal spinning is that if a crack forms or the object is dented, it must be scrapped. Repairing the object is not cost-effective.

METAL CUTTING

Metal cutting is "the process of removing unwanted material in the form of chips, from a block of metal, using cutting tool". A person who specializes in machining is called a machinist. A room, building or company where machining is done is called a Machine Shop.

The basic elements involved in this process are:

- A block of metal (work piece).
- Cutting Tool.
- Machine Tool.

- Cutting Fluid.

- Cutting speed (Primary Motion).

- Feed (Secondary Motion).

- Chips.

- Work holding and Fixturing.

- Force and Energy Dissipated, and

- Surface Finish.

The essential conditions for successful metal cutting are:

- Relative motion between work and cutting tool.

- Tool material must be harder than work material.

- Work and tool must be rigidly held by jig and fixtures.

- Sharp Cutting edge of cutting tool.

- Primary Motion (Cutting Speed).

- Secondary Motion (Cutting Feed).

Almost all the products produced by metal removal process, either directly or indirectly. The major disadvantages of the process are loss of material in the form of chips.

Types of Processes

Machining is not just one process; it is a group of processes. There are many kinds of machining operations. Each of which is specialized to generate a certain part geometry and surface finish quality.

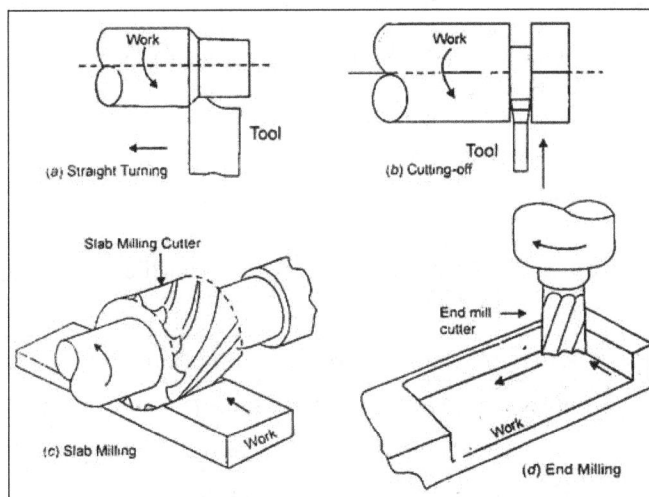

Common Cutting Processes.

Some of the more common cutting processes:

(i) Turning:

Turning is used to generate a cylindrical shape. In this process, the work piece is rotated and cutting tool removes the unwanted material in the form of chips. The cutting tool has single cutting edge. The speed motion is provided by the rotating work piece, and the feed motion is achieved by the cutting tool moving slowly in a direction parallel to the axis of rotation of the work piece.

(ii) Drilling:

Drilling is used to create a round hole. In this process, the cutting tool is rotated and feed against the work piece fixed in a holding device. The cutting tool typically has two or more cutting edges. The tool is fed in a direction parallel to its axis of rotation into the work piece to form the round hole.

(iii) Boring:

Boring is used to enlarge an already drilled hole. It is a fine finishing operation used in the final stage of product manufacture.

(iv) Milling:

Milling is used to remove a layer of material from the work surface. It is also used to produce a cavity in the work surface. In the first case it is known as slab-milling and in second case it is known as end- milling. Basically, the milling process is used to produce a plane or straight surface. The cutting tool used has multiple cutting edges. The speed motion is provided by the rotating milling cutter. The direction of the feed motion is perpendicular to the tool's axis of rotation.

(v) Cutting-off:

Cutting-off is used to cut the metal into two parts. In this operation, the work piece is rotated and cutting tool moves radially inward to separates the components.

Factors Influencing Metal Cutting Process

Various factors or parameters that affects the cutting process and so surface finish and accuracy of part geometry.

Table: Factors affecting cutting process.

S.NO	Parameter	Influence
1.	Cutting Parameters	Energy, force, power, temperature rise, tool life, type of chips, surface finish.
2.	Cutting fluid	Surface finish, dissipation.
3.	Tool geometry	Type of chips, chip flow direction, cutting force.
4.	Tool wear	Dimensional accuracy, surface finish, temperature rise, force and power requirement.
5.	Temperature rise	Thermal damage of workpart, dimensional accuracy, tool life, tool wear.
6.	Machinability	Tool life, force and power, surface finish.

7.	Continuous chips	Good surface finish, steady cutting forces.
8.	Built-up edge chips	Poor surface finish, thin stable built-up edge can protect tool surface.
9.	Dis-continuous chips	Fluctuating cutting forces,desirable for ease of chip disposal, vibrations and chatter.

Independent Variables

The major independent variables are:

- Cutting tool material, shape, geometry, angles.

- Work piece material, condition, temperature.

- Cutting parameters, such as speed, feed, and depth of cut.

- Cutting fluids.

- Machine tool specifications.

- Work holding and Fixturing.

Dependent Variables

Dependent variables are influenced by changes in independent variables.

The major dependent variables are:

- Types of chips formed.

- Temperature zone at work tool interface.

- Tool wear and failures.

- Surface finish.

- Force and energy in cutting process.

Methods of Metal Cutting

There are two basic methods of metal cutting based on cutting edge and direction of relative motion between tool and work:

- Orthogonal cutting process (Two Dimensional).

- Oblique cutting process (Three Dimensional).

Orthogonal Cutting Process

In orthogonal cutting process, the cutting edge is perpendicular (90 degree) to the direction of feed. The chip flows in a direction normal to cutting edge of the tool. A perfectly sharp tool will cut the metal on rack surface.

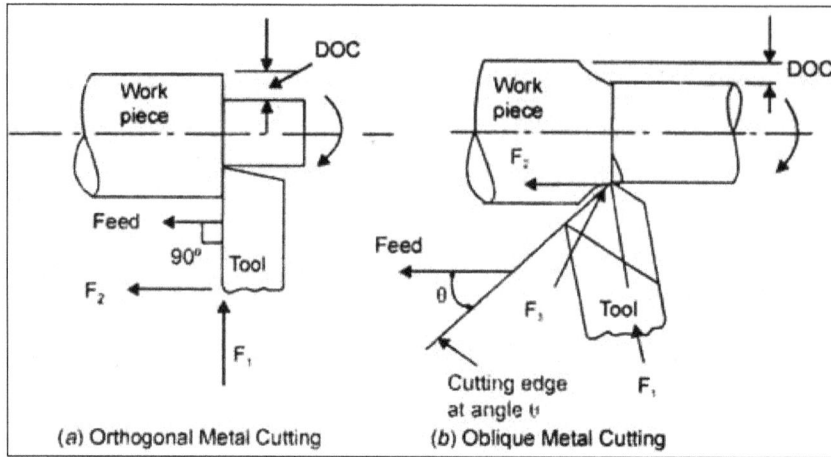

Types of Metal Cutting Process.

Oblique Cutting Process

In oblique cutting process, the cutting edge is inclined at an acute angle (less than 90 degree) to the direction of feed. The chip flows sideway in a long curl. The chip flows in a direction at an angle with normal to the cutting edge of the tool.

Table: Comparative features of orthgonal cutting and oblique cutting process.

S.No.	Basis	Orthogonal cutting process	Oblique cutting process
1.	2D/3D cutting process	It is a two-dimensional cutting process.	It is a three-dimensional cutting process.
2.	Cutting edge and direction of feed	The cutting edge of the tool remains at 90 degree to the direction of feed.	The cutting edge of the tool remains inclined at an acute angle to the direction of feed.
3.	Direction of chip flow	The chip flows in a direction normal to the cutting edge of the tool.	The chip flows in a direction at an angle with normal to the cutting edge of the tool.
4.	Components of cutting forces	There are only two-mutually perpen- dicular components of cutting forces acting on the tool.	There are three-mutually perpendicular components of cutting forces acting on the tool.
5.	Heat developed due to friction	During the process, the heat developed due to friction per unit area is more.	During the process, the heat developed due to friction per unit area is less.
6.	Tool life	Thus, tool life is smaller than obtained in oblique cutting (for same cutting conditions).	Thus, tools have longer life.
7.	Shear force	The shear force acts on a smaller area.	The shear force acts on a larger area of the cutting edge.
8.	Amount of metal removed	Results in less metal removal compared to oblique cutting.	Results in faster and more metal removal compared to orthogonal cutting.
9.	Relation between cutting edge and width of cut	The cutting edge is bigger than the width of cut.	The cutting edge is smaller than the width of cut.

10.	Figures	Cutting Tool / A / Chip / 90° / B / Work Piece / AB – Cutting Edge	Chip-flow angle / Cutting tool / A / Chip / F / B / <90° / Workpiece / Cutting edge inclination
		Orthogonal Cutting.	Oblique Cutting.

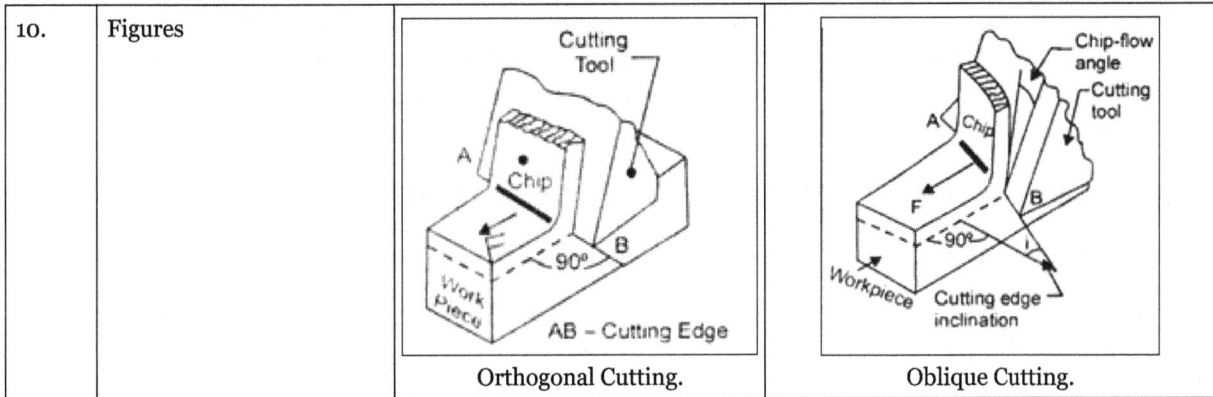

Principle of Metal Cutting

In this process, a wedge shaped tool moves relative to the work piece at an angle a. As the tool makes contact with the metal, it exerts pressure on it. Due to the pressure exerted by the tool tip, metal will shear in the form of chips on the shear plane AB. A chip is produced ahead of the cutting tool by deforming and shearing the material continuously, along the shear plane AB.

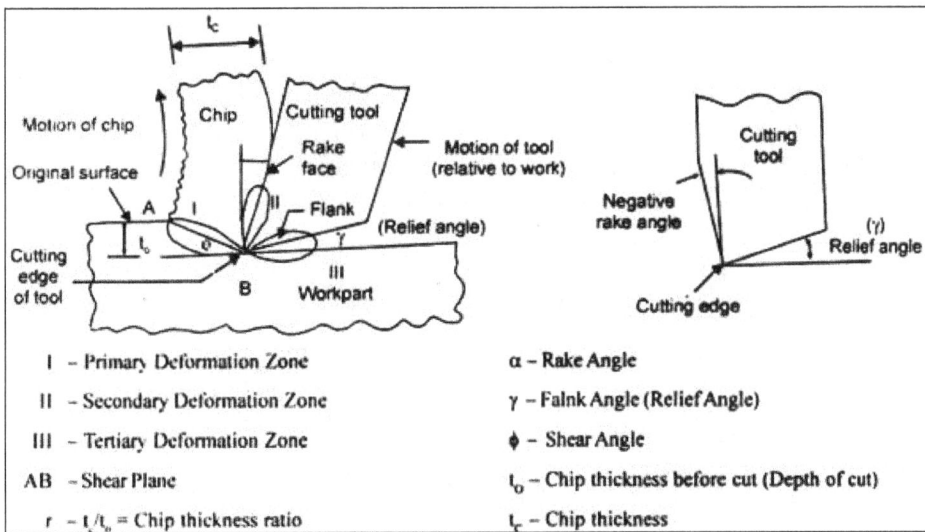

Principle of Metal Cutting.

The shear plane is actually a narrow zone and extends from the cutting edge of the tool to the surface of the work piece. The cutting edge of the tool is formed by two intersecting surfaces.

A detailed about various terminologies is given below:

(i) Rack Surface: It is the surface between chip and top surface of the cutting tool. It is the surface along which the chip moves upwards.

(ii) Flank Surface: It is the surface between work piece and bottom of the cutting tool. This surface is provided to avoid rubbing with the machined surface.

(iii) Rack Angle (α): It is the angle between the rack surface and the normal to work piece. Rack angle may be positive or negative.

(iv) Flank Angle/Clearance Angle/Relief angle (γ): It is the angle between the flank surface and the horizontal machined surface. It is provided for some clearance between flank surface and machined surface of work piece to avoid rubbing action of cutting tool to the finished surface.

(v) Primary Deformation Zone: It is the zone between tool tip and shear plane AB.

(vi) Secondary Deformation Zone: It is the zone between rack surface of the tool and chip.

(vii) Tertiary Deformation Zone: It is the zone between flank surface of the tool and machined surface of work piece.

Almost all the cutting processes involve the same shear-deformation theory. The cutting tool used in cutting process may be single-point or multi-point cutting tool. Turning, threading, and shaping, boring, chamfering, and facing are some cutting operations done by single point cutting tool. Milling, drilling, grinding, reaming and broaching are some cutting operations done by multi-point cutting tool.

Mechanics of Chip Formation

A typical metal cutting process by single point cutting tool is shown in fig. In this process a wedge shaped tool moves relative to the work piece at an angle α. As the tool makes contact with the metal, it exerts pressure on it. Due to the pressure exerted by the tool tip, metal will shear in the form of chips on the shear plane AB. A chip is produced ahead of the cutting tool by deforming and shearing the material continuously along the shear plane AB.

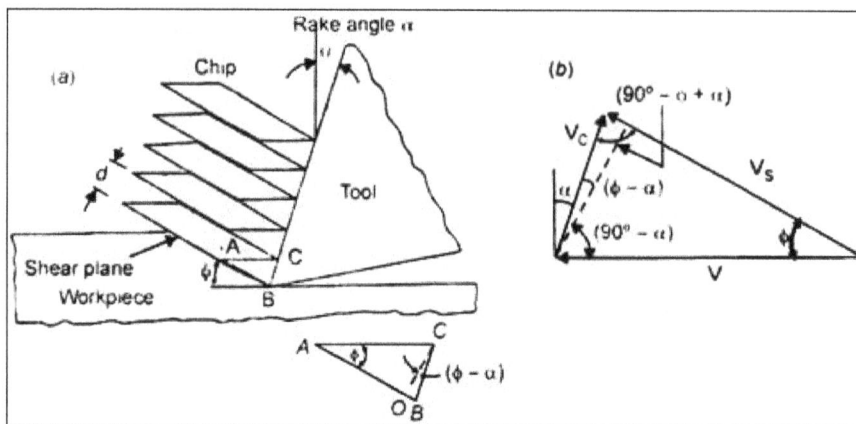

Mechanics of Chip Formation.

Microscopic study shows that chips are produced by the shearing process. Shearing process in chip formation is similar to the motion of cards in a deck sliding against each other, Shearing takes place along a shear zone (shear plane). The shear plane is actuality a narrow zone. It extends from the cutting edge of the tool to the surface of the work piece.

This plane is at an angle called the shear angle (φ), with the surface of the work piece. Shear zone has a major influence on the quality of the machined surface. Below the shear plane the work piece is under formed while above the shear plane the chip already formed and moving upwards to the tool face.

The ratio of thickness of chip before cut (t_o) to the thickness of chip after cut (t_c) is known as chip thickness ratio.

It is generally represented by r, which can be expressed as:

$$r = \frac{t_0}{t_c} = \frac{\sin\phi}{\cos(\phi - \alpha)}$$

The chip thickness after cut (t_c) is always greater than the chip thickness before cut (t_o). Therefore, the value of r is always less than unity. The reciprocal of r is known as chip compression ratio or chip reduction ratio $(1/r)$. The chip reduction ratio is a measure of how thick the chip has become compared to the depth of cut (t_o). Thus the chip reduction ratio is always greater than unity.

Derivation to Calculate Shear Angles

Considering orthogonal cutting process to derive the expression to calculate shear angle, The cutting tool is defined by rake angle (α) and clearance or relief angle (γ). The chip is formed perpendicular to the cutting edge of the tool.

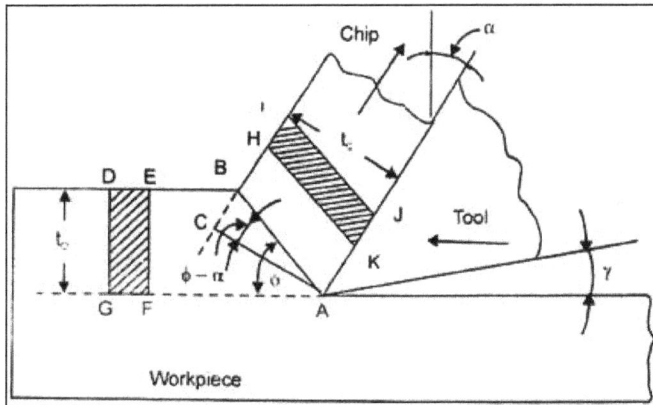

Mechanics of Chip Formation in Orthogonal Cutting.

Following are some assumptions made to the mechanics of chip formations:

(i) Tool should contacts the chip on its rake face.

(ii) Plain strain conditions considered. It means there is no side flow of the chip during cutting.

(iii) The deformation zone is very thin (in the order of 10^{-2} to 10^{-3} mm) adjacent to the shear plane AB.

Where,

α – Rake angle,

γ – Clearance (relief) angle,

φ – Shear angle,

AB – Shear plane,

t_o – Uncut chip thickness,

t_c – Chip thickness (deformed),

Area DEFG – Area of uncut chip,

Area HIJK – Chip area after cutting.

From figure,

$$t_o = AB \sin \phi$$
$$t_c = AB \cos (\phi - \alpha)$$

The chip thickness ratio can be expressed as,

$$r = \frac{t_o}{t_c}$$

The chip reduction ratio becomes,

$$\frac{1}{r} = \frac{t_c}{t_o}$$

$$\frac{1}{r} = \frac{AB \cos (\phi - \alpha)}{AB \sin \phi}$$

$$\frac{1}{r} = \frac{\cos (\phi - \alpha)}{\sin \phi}$$

$$\frac{1}{r} = \frac{\cos \phi \cos \alpha + \sin \phi \sin \alpha}{\sin \phi}$$

$$\frac{1}{r} = \cos \phi \cos \alpha + \sin \alpha$$

$$\cos \phi = \frac{\left(\dfrac{1}{r} - \sin \alpha\right)}{\cos \alpha} = \frac{1 - r \sin \alpha}{r \cos \alpha}$$

$$\tan \phi = \frac{r \cos \alpha}{1 - r \sin \alpha}$$

This is the required relation to calculate the shear angle (φ). This relation shows that φ depends upon the t_o, t_c, and α (rake angle). It means by measuring t_o, t_c and a of the tool, shear angle (φ) can be determined using expression.

The chip thickness ratio (r) may be determined by following methods:

- By using continuity equation.

- By weighing a known length of chip.

- By knowing chip velocity (V_c) and work piece velocity (V).

By using Continuity Equation

Original weight of chip before cut = weight of chip after cut.

$$W_b = W_a$$

If, l_1, b_1, t_o = length, breadth and thickness of chip before cut.

and l_2, b_2, t_c = length, breadth and thickness of chip after cut.

$$\rho = \text{density of material.}$$

Then,

$$W_b = W_a$$

$$\rho\left(l_1, b_1, t_o\right) = \rho l_2, b_2, t_c$$

and also by assumption

$$b = b_c$$

or

$$b_1 = b_2 \quad \left(\text{No side flow of chip}\right)$$

Then,

$$l_1 t_o = l_2 t_c$$

or

$$\frac{t_o}{t_c} = \frac{l_2}{l_1}$$

or

$$r = \frac{t_o}{t_c} = \frac{l_2}{l_1}$$

By Weighing a known Length of Chip

If the length of cut is not directly known then we can estimate by weighing a known length of chip; then

Weight of chip $\qquad W_C = \rho l_2 b t_c = \rho l b t$

Then, $\qquad l = \dfrac{W_C}{\rho b t}$

the calculate 'r' and ϕ from above equations.

(iii) By knowing Chip Velocity (V_c) and Work Piece Velocity (V):

Applying continuity equation as:

$$V \rho b_1 t_o = V_c \rho b_2 t_c$$

$$V \rho b_1 t_o = V_c \rho b_2 t_c$$
$$V t_o = V_c t_c$$
$$\frac{t_0}{t_c} = \frac{V_c}{V} = r$$

By putting the value of r and α, we can obtained the shear angle (φ).

Velocities in Metal Cutting Process

Because of relative motion between tool tip and work piece and chip removed, there are three types of velocities comes into existence.

These are following:

- Cutting Speed or Velocity (V): It is the velocity of the cutting tool relative to the work piece.

- Shear Velocity (V_s): It is the velocity of chip relative to the work piece. In other way, the velocity at which shearing takes place.

- Chip Velocity (V_c): It is the velocity of the chip up the tool face (rake face) during cutting.

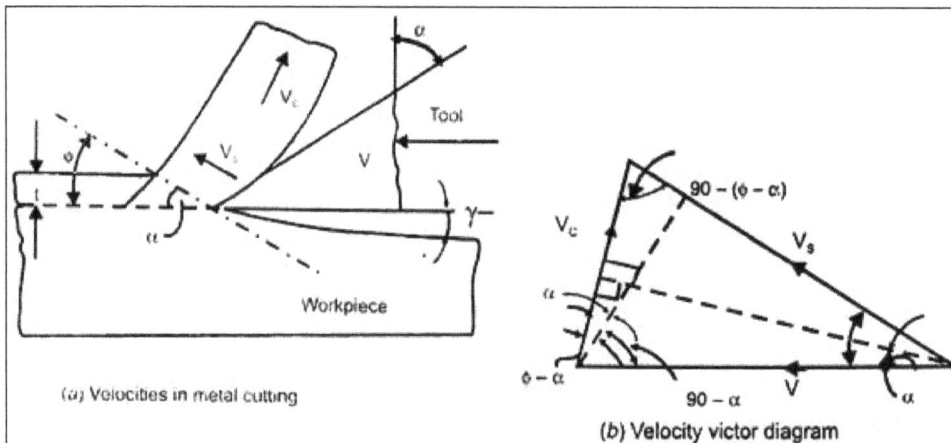

(a) Velocities in metal cutting
(b) Velocity victor diagram

Velocities Metal Cutting Process.

Let,   V – Cutting Velocity

Vs – Shear Velocity

Vc – Chip velocity

φ – Shear angle

α – Rake angle

r – Chip thickness ratio

γ – Clearance angle

Using continuity equation, the volume of metal removal before and after is same, therefore:

Vt = Vc tc

Vc/V = t/tc = r

Using sine rule to the velocity vectors we can write:

$$\frac{V_s}{V} = \frac{\cos\alpha}{\cos(\phi-\alpha)}$$

$$\therefore \qquad V_s = \frac{V\cos\alpha}{\cos(\phi-\alpha)}$$

Similarly, $\qquad V_c = \frac{V\sin\phi}{\cos(\phi-\alpha)}$

and also, $\qquad V_c = \frac{V}{r}$

From kinematics theory, the relative velocity of two bodies (tool and chip) is equal to vector difference between their velocities relative to reference body (work piece), then:

V = VC + VS

Forces Acting on the Chip

(i) Shear Force (F_s): It is acting along shear plane. It is the resistance to shear of metal.

(ii) Normal Force (F_n): It is perpendicular to the shear plane generated by the work piece.

(iii) Normal Force (N): It is exerted by the tool tip on the chip.

(iv) Fractional Resistance Force (F): It is acting on the chip and it acts against the chip motion along the tool face.

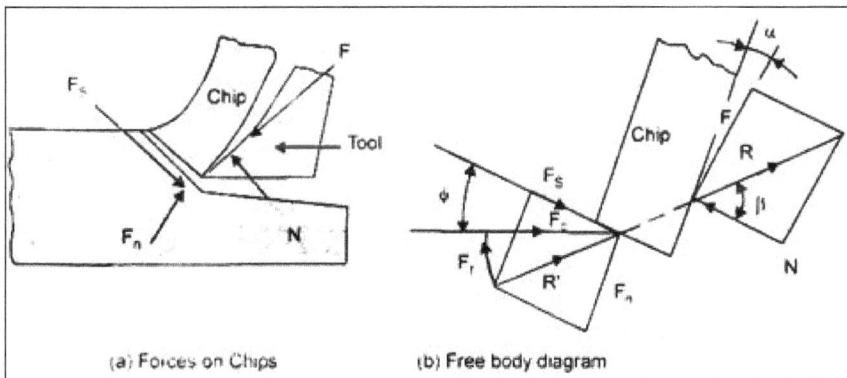

(a) Forces on Chips   (b) Free body diagram

Forces Action on the Chip.

Figure indicates the free body diagram of chip which is in equilibrium under the action of resultant forces equal and opposite in magnitude and direction.

Thus,

$$\vec{R} = \vec{F} + \vec{N} \text{ and } \vec{R}' = \vec{F}_s + \vec{F}_n$$

Since, the chip is in equilibrium condition, so we can say that:

$$\vec{R} = -\vec{R}'$$

Types of Chips Produced in Machining

The chips produced in metal cutting process are not alike. The type of chip produced depends on the material being machined and the cutting conditions.

These conditions include:

- Type of cutting tool used.
- Speed and rate of cutting.
- Tool geometry and cutting angles.
- Condition of machine.
- Presence/Absence of cutting fluid, etc.

The study of chips produced are very important because the type of chips produced influence the surface finish of the work piece, tool life, vibrations, chatter, force and power requirements, etc.

It is Important to note that a Chip has two Surfaces:

- Shiny Surface: It is the surface which is in contact with the rake face of the tool. Its shiny appearance is caused by the rubbing of the chip as it moves up the tool face.
- Rough Surface: It is the surface which does not come into contact with any solid body. It is the original surface of the work piece. Its rough appearance is caused by the shearing action.

Types of Commonly Observed Chips in Practice

These are discussed below:

- Continuous chips.
- Continuous chips with built-up edge.
- Discontinuous or Segmental chips.

Continuous Chips

Continuous chips are produced when machining more ductile materials such as mild steel, copper and aluminum.

Due to large plastic deformation possible with more ductile materials, longer continuous chips are produced. It is associated with good tool angles, correct speeds and feeds, and the use of cutting fluids.

Advantages

- They generally produce good surface finish.

- They are most desirable because the forces are stable and operation becomes vibration less.

- They provide high cutting speeds.

Disadvantages

- Continuous chips are difficult to handle and dispose off.

- Continuous chips coil in a helix and curl around the tool and work and even may injure operator if sudden break loose.

- Continuous chips remain in contact with the tool face for a longer period, resulting in more frictional heat is used to break the continuous chip into small sections so that the chips cannot curl around the cutting tool.

The simplest form of chip breaker is made by grinding a groove on the tool face a few millimeters behind the cutting edge. Sometimes, a small metal plate stick with cutting tool face is used as a chip breaker.

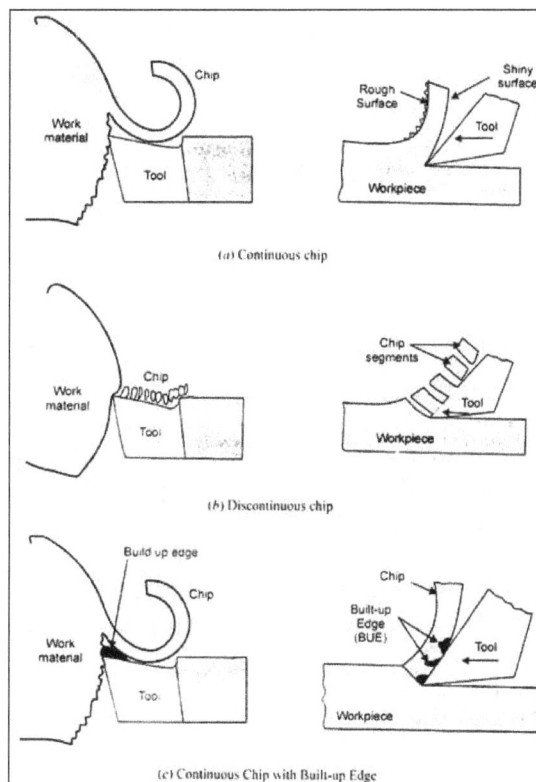

Types of Chips Produced in Metal Cutting.

Favourable Cutting Conditions

The favourable cutting conditions for production of continuous chips are following:

- Machining more ductile materials such as copper, aluminum.

- High cutting speed with fine feed.

- Larger rake angle.

- Sharper cutting edge.

- Efficient lubricant.

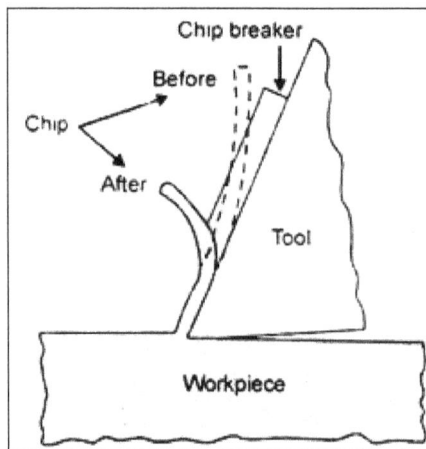

Action of Chip Breaker.

Continuous Chips with Built-up Edge

Continuous chips with Built-up Edge (BUE) are produced when machining ductile materials under following conditions:

- High local temperature in cutting zone.

- Extreme pressure in cutting zone.

- High friction at tool-chip interface.

The above machining conditions cause the work material to adhere or stick to the cutting edge of the tool and form Built-Up Edge (BUE). The built-up edge generates localized heat and friction, resulting in poor surface finish, power loss.

The built-up edge is commonly observed in practice. The built-up edge changes its size during the cutting operation. It first increases, then decreases, and then again increases, etc. This cycle is source of vibration and poor surface finish.

Advantages

Although built-up edge is generally undesirable, a thin, stable BUE is usually desirable because it reduces wear by protecting the rake face of the tool.

Disadvantages

- This is a chip to be avoided.

- The phenomenon results in a poor surface finish and damage of the tool.

Favourable Cutting Conditions

The favourable cutting conditions for production of continuous chips with built-up edge are following:

- Low cutting speed.

- Low rake angle.

- High feed.

- Inadequate supply of coolant.

- Higher affinity (tendency to form bond) of tool material and work material.

Reduction or Elimination of BUE

The tendency to form BUE can be reduced or eliminate by any one of the following practices:

- Increasing the cutting speed.

- Increasing the rake angle.

- Decreasing the depth of cut.

- Using an effective cutting fluid.

- Using a sharp tool.

- Light cuts at higher speeds.

Discontinuous or Segmental Chips

Discontinuous chips are produced when machining more brittle materials such as grey cast iron, bronze, brass, etc. with small rake angles. These materials lack the ductility necessary for appreciable plastic chips deformation. The material fails in a brittle fracture ahead of the tool edge along the shear zone. This results in small segments of discontinuous chips. There is nothing wrong with this type of chip in these circumstances.

Advantages

- Since the chips break-up into small segments, the friction between the tool and the chip reduces, resulting in better surface finish.

- These chips are convenient to collect, handle and dispose of.

Disadvantages

- Because of the discontinuous nature of chip formation, forces continuously vary during cutting process.

- More rigidity or stiffness of the cutting tool, holder, and work holding device is required due to varying cutting forces.

- Consequently, if the stiffness is not enough, the machine tool may begin to vibrate and chatter. This, in turn, adversely affects the surface finish and accuracy of the component. It may damage the cutting tool or cause excessive wear.

Favourable Cutting Conditions

The Favourable Cutting Conditions for Production of Discontinuous Chips are following:

- Machining brittle materials.
- Small rake angles.
- Very low cutting speeds.
- Low stiffness of machine tool.
- Higher depth of cuts.
- Inadequate lubricant.
- Materials that contain hard inclusions and impurities.

METAL JOINING

Metal joining is a controlled process used to fuse metals. There are several techniques of metal joining of which welding is one of the more basic forms. Expertise and technological advances have enabled metal joining innovations, which in turn have lead to component advances in many industrial sectors, including aerospace. Metal joining includes specialised processes such as electron beam welding, vacuum, and honeycomb brazing – complex operations requiring a fusion of expertise and technology.

Method

Soldering

There are various common methods of joining parts together used in Engineering works. Soldering is one of the techniques of the joining method, but thinner parts.

Solder

Solder is an Alloy of Tin and Lead, sometime Antimony and Bismuth are also added.

Types of Soldering

There are three types of soldering:

- Soft Soldering;
- Hard Soldering;
- Brazing.

Soft Soldering

Soft Soldering is a processes of joining two or more pieces of similar or dissimilar thinner parts by an alloy called 'Solder', which has a lower melting point than the base metals. The temperature required is approx. 300°C to 350°C for Soft Soldering.

Hard Soldering

In this process is required an extra external heat which comes from Blow-Lamp or Oxy-acetylene flame to melt the solder. Generally it is used for underground cable-joint of Telephone, Electric, etc.

Brazing

Brazing is a kind of Hard Soldering. In this process, heat is applied from Oxy-acetylene flame. But in Soft Soldering no extra heat is required.

Table: Types of soft soldering and its composition.

Sl. No.	Name of Soldering	Tin Point	Lead	Antimony	Bismuth	Melting Point
1.	Electrical	95%	5%	x	x	220 °C
2.	Tinsmith	60%	40%	X	X	192 °C
3.	Ordinary	50%	50%	X	X	205 °C
4.	Plumber	33%	66%	1%	X	205 °C
5.	Fine	66%	33%	X	1%	171 °C

Equipment's and Jointing Process of Soft Soldering

Equipment's required for Soldering are:

- Soldering Iron.
- Open Hearth Furnace.
- Electrical Soldering Iron.
- Zinc Chloride Solution or Acid, Emery Paper, etc.

How to Join by Solder

One point should be remembered—only thinner parts can be soldered by this process. A thin sheet, like Galvanized sheet, may be joined to another Mechanical process by Riveting and by Soldering to Soft Soldering. Before Soldering, the parts should be cleaned with ACID (Zinc Chloride solution) or liquid Fluxes to remove dust, oil and grease, and other foreign metals.

The Soft Soldering Iron is heated by an open hearth furnace or by electrically. Thinner parts of Tin, Copper, Brass, Aluminium, etc. and electronic parts like Television and electrical goods can be repaired by this process.

Brazing

Brazing may be defined as "Hard Soldering". Brazing is a method of joining metals by the use of a fusible alloy consisting largely of Brass, and melting at a temperature above 600°C. Brazing produces a much stronger joint than soft soldering but requires greater heat from Oxy-acetylene flame. But it should be remembered that it is a temporary joint. It can be easily separated after heating of the welded bronze metal.

In this process, non-ferrous or alloy filler metal is used whose melting point is higher— about 1,000°F (540°C)—but lower than that of the metals being welded. The Bronze or Brass filler rod should be coated with a flux of a deoxidizing agent. This operation is called "fluxing the rod".

Moreover, the operation Tinning is very important in Bronze welding or Brazing. By moving the flame around the starting point of the weld, the base metal gradually becomes red hot. Tinning is an operation in which a molecular union between the Bronze filler metal and the base metal is achieved. The strength of weld mainly depends upon the molecular union.

The commonly used Brazing in modern industries are:

- Torch Brazing.
- Hearth Brazing.
- Furnace Brazing.
- Resistance Brazing.
- Induction Brazing.
- Dip Brazing.
- Salt bath Brazing.
- Twin-carbon-arc Brazing.

Brazing Flux

Flux is a chemical compound or mixture of deoxidizing agent used as powder, paste, liquid, granular, and gaseous. The flux employed in Brazing operations depends entirely upon the type of operation and the Brazing alloy being employed.

When metals are heated in contact with air, oxygen from oxides cause poor quality low strength welds, or, in some cases, may even make welding impossible. For this reason it is generally desirable to add a flux to the welding area— this being a substance capable of dissolving the oxide.

It is important to know that "The ordinary grades of mild steel provide an exception to the rule, because they may be successfully welded without the use of a flux. The reason is that mild steel contains considerably, more silicon and manganese which act as deoxidizing and fluxing agents".

No single flux is suitable for all metals. So it is necessary to choose a flux developed specially for the particular metal being welded. For ferrous materials, borax, sodium carbonate, sodium bicarbonate, and sodium silicate have been found to give excellent results, together with small additions of vigorous deoxidizing substances.

The borax forms compound with iron oxide whilst the carbonate is a cleanser and promotes fluidity.

For copper and copper alloy mixtures of sodium and potassium borates, carbonates, chlorides, sulphate, and boric acid have been found suitable for removing cuprous oxide, thus preventing mechanically unsafe welds. Flux for aluminium consists of alkaline fluorides, chlorides, and bi-sulphates.

For magnesium alloy, the fluxes are similar in composition to those used for aluminium and its alloys. All fluxes are chemically active and very corrosive. It is, therefore, essential to remove all traces of flux from the finished weld. The flux residue may be removed by recovering the weld repeatedly with hot water, steam or by the use of picking baths.

Purposes of Flux

Firstly, Flux serves several purposes. Flux acts as a good insulator and concentrates heat in a relatively small welding zone—thus improving the fusion of the welding rod and the melted parent metal.

Secondly, the molten portion of the flux floats as a liquid blanket over the molten pool of the electrode and parent metal, protecting it from the atmosphere and reducing to a minimum its pick-up of Oxygen and Nitrogen.

Thirdly, the flux acts as a cleanser for the weld metal, absorbing impurities and adding alloying elements such as manganese and silicon. Consequently, the weld metal is clean and has excellent physical properties.

Types of Fluxes

The three types of flux:

1. Borax Flux: Borax flux is employed with the copper zinc Brazing alloys. It has a melting point of approximately 750°C. The correct type of Borax to use is fused borax — Borax glass — to which, in some instances, boric may be added.

Common borax, which contains about 50% by weight of water of crystallization, should not be used, as when the powder is heated it swells considerably and may be blown away from the joint by the force of the flame used for welding.

2. Fluoride: This type of flux is employed with the silver solder alloys. Its melting point varies from 600°C to 750°C. It can be made up in the Brazing shop but it is generally advisable to purchase the recommended flux from the makers of one or other of the proprietary silver solder alloys.

3. Mixed Halide: This type of flux is employed for the Brazing of aluminium alloys. It is used for Brazing for several functions; the most important is to protect the surface of the work from oxidation during heating. Secondly, to dissolve any metallic oxides which may already be present or may be formed during the Brazing operation.

The flux should also serve to reduce the surface tension of the molten brazing alloy in relation to the metal on which it is required to flow, enabling the molten metal to wet the surface to be jointed.

So, by use of a flux, welding is made as easy and free from difficulties as possible. Fluxes are chemical compounds used to prevent oxidation and other unwanted chemical reactions. They help to make the welding process easier and ensure a good and sound weld.

Riveting

Rivets are used when it is required to connect permanently two or more pieces of heavy section metal. Light gauge sheet can also be joined by Riveting. It is required specially in heavy sections such as, Bridge, Wagon, Tanker, Boiler, etc. In engineering rivets are usually made of wrought iron or mild steel.

A rivet consists essentially of three parts—the head, the two types of rivets:

1. Snap head rivet.

2. Pan head rivet.

Process of Riveting

There are four processes of riveting:

1. Single riveting.

2. Double.

3. Chain riveting.

4. Zig-Zag riveting.

The holes in the plates are either punched (light gauge sheet) or drilled. Punching is liable to damage the plates, especially when they are of hard wrought iron. On the best work, plates are 'tacked' or clamped together by a few bolts here and there along the seam.

So the remaining holes may be drilled through both plates without any danger of trouble arising through holes being out of alignment. A Burr is usually left after every drilling operation as a riveted joint may aid shear.

After riveting, when they cool, rivets tend to pull the plates more closely together. They cannot be removed without chipping off the heads.

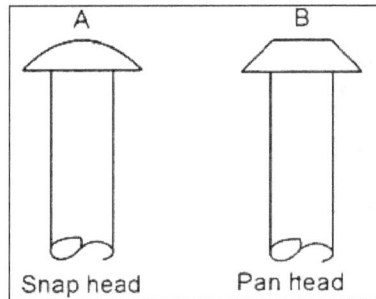

Welding

Welding is a permanent joint. Welding is seen everywhere—No welding, No industry. It is an industrial art of joining metals by pressure after heating to a plastic state or joining two pieces of metals by fusion or by pressure to form a solid union or a compact mass.

Group of Welding

Mainly two groups of welding are used in industry:

- Autogenous Welding;

- Heterogeneous Welding.

The first type of welding may be of similar metals; and the second type may be of similar or dissimilar metals to be welded with the third additional metal or alloy which has got low melting temperature than the metal to be welded. This is Autogenous welding.

Welding Position

Welding positions are of five types:

- Flat and Down hand or Ground position;

- Inclined position;

- Horizontal position;

- Vertical position; and

- Overhead position.

Weld Slope

It is the angle formed by the line of weld root and the Horizontal reference plate.

Weld Rotation

It is the angle formed between the upper position of a vertical reference plane passing through the line of weld root and the part of a plane passing through the weld root and the point on the face of the weld equidistant from both edges of the weld.

Welding Parameters

Before welding, some causes of action or Parameters are to be followed:

- Classification and size of electrodes.

- Current and open circuit voltage.

- Length of runs or passes per electrode of speed of travel.

- Number and arrangement of runs in multi-run weld.

- Position of welding.

- Preparation and set-up of parts.

- Welding sequence.

- Preheating and post-heating.

METAL CASTING

Metal casting is defined as the process in which molten metal is poured into a mold that contains a hollow cavity of a desired geometrical shape and allowed to cool down to form a solidified part.

A mold is formed into the geometric shape of a desired part. Molten metal is then poured into the mold, the mold holds this material in shape as it solidifies. A metal casting is created. Although this seems rather simple, the manufacturing process of metal casting is both a science and an art An open mold is a container, like a cup, that has only the shape of the desired part. The molten material is poured directly into the mold cavity which is exposed to the open environment.

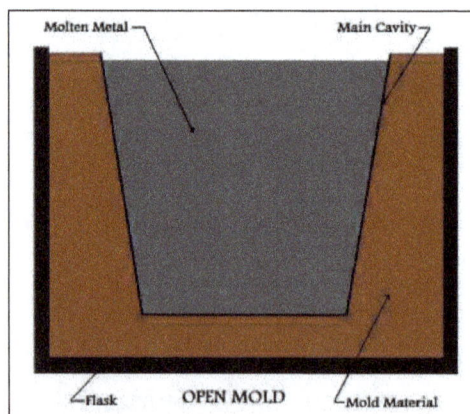

This type of mold is rarely used in manufacturing production, particularly for metal castings of any level of quality. The other type of mold is a closed mold, it contains a delivery system for the molten material to reach the mold cavity, where the part will harden within the mold. A very simple closed mold. The closed mold is, by far, more important in manufacturing metal casting operations.

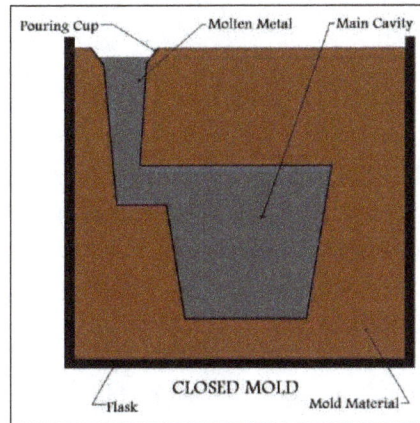

There are many different metal casting processes used in the manufacture of parts. Two main branches of methods can be distinguished by the basic nature of the mold they employ. There is expendable mold casting and permanent mold casting. As the name implies, expendable molds are used for only one metal casting while permanent molds are used for many. When considering manufacturing processes, there are advantages and disadvantages to both.

Expendable Mold	Permanent Mold
• Can produce one metal casting only.	• Can manufacture many metal castings.
• Made of sand, plaster, or other similar material.	• Usually made of metal or sometimes a refractory ceramic.
• Binders used to help material hold its form.	
• Mold that metal solidifies in must be destroyed to remove casting.	• Mold has sections that can open or close, permitting removal of the casting.
• More intricate geometries are possible for casting.	• Need to open mold limits part shapes.

Patterns

Expendable molds require some sort of pattern. The interior cavities of the mold, in which the molten metal will solidify, are formed by the impression of this pattern. Pattern design is crucial to success in manufacture by expendable mold metal casting. The pattern is a geometric replica of the metal casting to be produced. It is made slightly oversize to compensate for the shrinkage that will occur in the metal during the casting's solidification, and whatever amount of material that will be machined off the cast part afterwards. Although machining will add an extra process to the manufacture of a part, machining can improve surface finish and part dimensions considerably. Also, increasing the machine finish allowance will help compensate for unknown variables in shrinkage, and reduce trouble from areas of the metal casting that may have been originally too thin or intricate.

Pattern Material

The material from which the pattern is made is dependent upon the type of mold and metal casting process, the casting's geometry and size, the dimensional accuracy required, and the number of metal castings to be manufactured using the pattern. Patterns can be made from wood, like pine (softwood), or mahogany (hardwood), various plastics, or metal, like aluminum, cast iron, or steel. In most manufacturing operations, patterns will be coated with a *parting agent* to ease their removal from the mold.

Cores

For metal castings with internal geometry *cores* are used. A core is a replica, (actually an inverse), of the internal features of the part to be cast. Like a pattern, the size of the core is designed to accommodate for shrinkage during the metal casting operation. Unlike a pattern, a core remains in the mold while the metal is being poured. Hence, a core is usually made of a similar material as the mold. Once the metal casting has hardened, the core is broken up and removed much like the mold. Depending upon the location and geometry of the core within the casting, it may require that it is supported during the operation to prevent it from moving or shifting. Structural supports that hold the core in place are called *chaplets*. The chaplets are made of a material with a higher melting temperature than the casting's material, and become assimilated into the part when it hardens. Note that when manufacturing a metal casting with a permanent mold process, the core will be a part of the mold itself.

The Mold

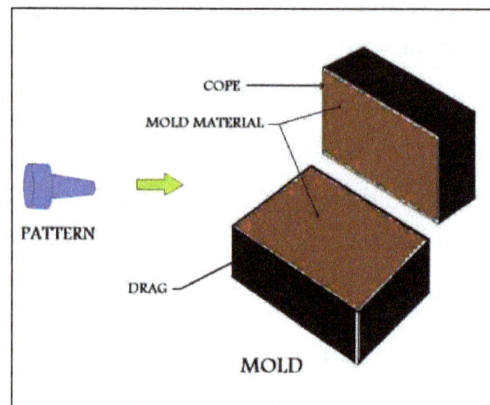

When manufacturing by metal casting, consideration of the mold is essential. The pattern is placed in the mold and the mold material is packed around it. The mold contains two parts, the drag (bottom), and the cope (top). The parting line between the cope and drag allows for the mold to be opened and the pattern to be removed once the impression has been made.

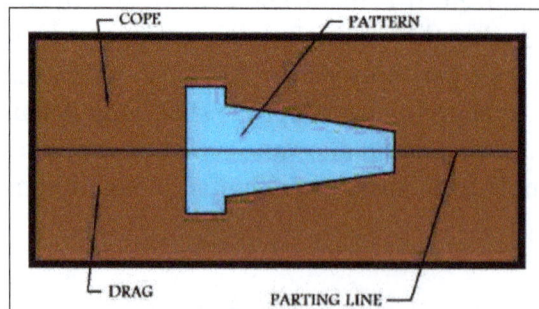

The core is placed in the metal casting after the removal of the pattern. Figure shows the pattern impression with the core in place.

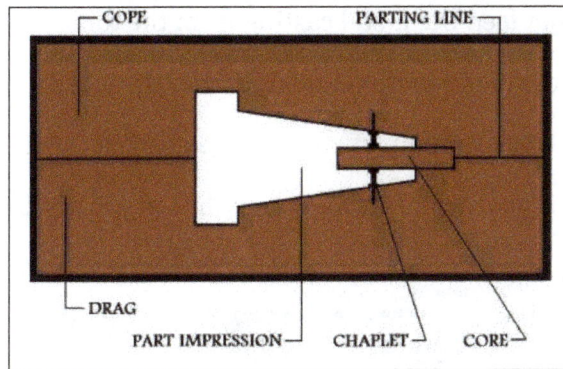

Now the impression in the mold contains all the geometry of the part to be cast. This metal casting setup, however, is not complete. In order for this mold to be functional to manufacture a casting, in addition to the impression of the part, the mold cavity will also need to include a gating system. Sometimes the gating system will be cut by hand or in more adept manufacturing procedures, the gating system will be incorporated into the pattern along with the part. Basically, a gating system functions during the metal casting operation to facilitate the flow of the molten material into the mold cavity.

Elements of a Gating System

Pouring Basin:

This is where the molten metal employed to manufacture the part enters the mold. The pouring basin should have a projection with a radius around it to reduce turbulence.

Down Sprue

From the pouring basin, the molten metal for the casting travels through the down sprue. This should be tapered so its cross-section is reduced as it goes downward.

Sprue Base

The down sprue ends at the sprue base. It is here that the casting's inner cavity begins.

- Ingate/Choke Area: Once at the sprue base, the molten material must pass through the ingate in order to enter the inner area of the mold. The ingate is very important for flow regulation during the metal casting operation.

- Runners: Runners are passages that distribute the liquid metal to the different areas inside the mold.

- Main Cavity: The impression of the actual part to be cast is often referred to as the main cavity.

- Vents: Vents help to assist in the escape of gases that are expelled from the molten metal during the solidification phase of the metal casting process.

- Risers: Risers are reservoirs of molten material. They feed this material to sections of the mold to compensate for shrinkage as the casting solidifies. There are different classifications for risers.

- Top Risers: Risers that feed the metal casting from the top.

- Side Risers: Risers that feed the metal casting from the side.

- Blind Risers: Risers that are completely contained within the mold.

- Open Risers: Risers that are open at the top to the outside environment.

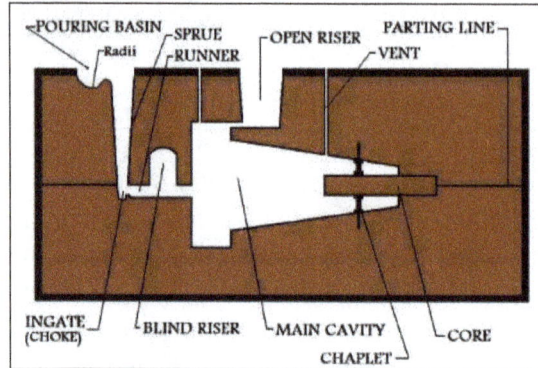

Gating System for Casting.

DEEP DRAWING

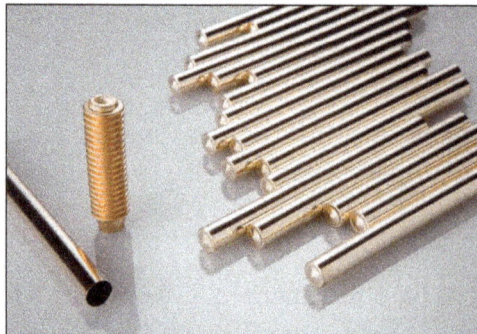

Example of deep drawn part.

Deep drawing is a sheet metal forming process in which a sheet metal blank is radially drawn into a forming die by the mechanical action of a punch. It is thus a shape transformation process with material retention. The process is considered "deep" drawing when the depth of the drawn part exceeds its diameter. This is achieved by redrawing the part through a series of dies. The flange region (sheet metal in the die shoulder area) experiences a radial drawing stress and a tangential compressive stress due to the material retention property. These compressive stresses (hoop stresses) result in flange wrinkles (wrinkles of the first order). Wrinkles can be prevented by using a blank holder, the function of which is to facilitate controlled material flow into the die radius.

Process

The total drawing load consists of the ideal forming load and an additional component to compensate for friction in the contacting areas of the flange region and bending forces as well as unbending forces at the die radius. The forming load is transferred from the punch radius through the drawn part wall into the deformation region (sheet metal flange). In the drawn part wall, which is in contact with the punch, the hoop strain is zero whereby the plane strain condition is reached. In reality, mostly the strain condition is only approximately plane. Due to tensile forces acting in the part wall, wall thinning is prominent and results in an uneven part wall thickness, such that the part wall thickness is lowest at the point where the part wall loses contact with the punch, i.e., at the punch radius.

The thinnest part thickness determines the maximum stress that can be transferred to the deformation zone. Due to material volume constancy, the flange thickens and results in blank holder contact at the outer boundary rather than on the entire surface. The maximum stress that can be safely transferred from the punch to the blank sets a limit on the maximum blank size (initial blank diameter in the case of rotationally symmetrical blanks). An indicator of material formability is the limiting drawing ratio (LDR), defined as the ratio of the maximum blank diameter that can be safely drawn into a cup without flange to the punch diameter. Determination of the LDR for complex components is difficult and hence the part is inspected for critical areas for which an approximation is possible. During severe deep drawing the material work hardens and it may be necessary to anneal the parts in controlled atmosphere ovens to restore the original elasticity of the material.

Commercial applications of this metal shaping process often involve complex geometries with straight sides and radii. In such a case, the term stamping is used in order to distinguish between the deep drawing (radial tension-tangential compression) and stretch-and-bend (along the straight sides) components. Deep drawing is always accompanied by other forming techniques within the press. These other forming methods include:

- Beading: Material is displaced to create a larger, or smaller, diameter ring of material beyond the original body diameter of a part, often used to create O-ring seats.

- Bottom Piercing: A round or shaped portion of metal is cut from the drawn part.

- Bulging: In the bulging process a portion of the part's diameter is forced to protrude from the surrounding geometry.

- Coining: Material is displaced to form specific shapes in the part. Typically coining should not exceed a depth of 30% of the material thickness.

- Curling: Metal is rolled under a curling die to create a rolled edge.

- Extruding: After a pilot hole is pierced, a larger diameter punch is pushed through, causing the metal to expand and grow in length.

- Ironing / Wall Thinning: Ironing is a process to reduce the wall thickness of parts. Typically ironing should not exceed a depth of 30% of the material thickness.

- Necking: A portion of the part is reduced in diameter to less than the major diameter.

- Notching: A notch is cut into the open end of the part. This notch can be round, square, or shaped.

- Rib Forming: Rib forming involves creating an inward or outward protruding rib during the drawing process.

- Side Piercing: Holes are pierced in the side wall of the drawn part. The holes may be round or shaped according to specifications.

- Stamping / Marking: This process is typically used to put identification on a part, such as a part number or supplier identification.

- Threading: Using a wheel and arbor, threads are formed into a part. In this way threaded parts can be produced within the stamping press.

- Trimming: In the Trimming process, excess metal that is necessary to draw the part is cut away from the finished part.

Often components are partially deep draw in order to create a series of diameters throughout the component. It common use to consider this process as a cost saving alternative to turned parts which require much more raw material.

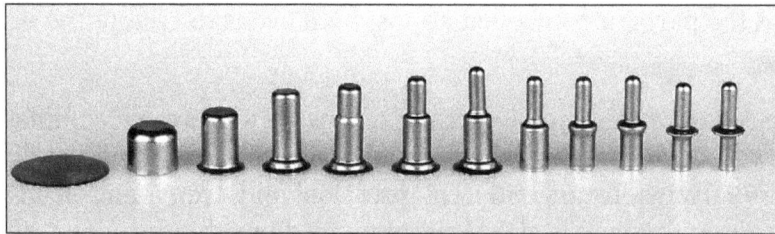
Example of deep drawn line.

The sequence of deep drawn components is referred to as a "deep draw line". The numbers of components that form the deep draw line is given by the quantity of "stations" available in the press. In the case of mechanical presses this is determined by the number of cams on the top shaft.

For high precision mass productions, it is always advisable to use a transfer press also known as eyelet press. The advantage of this type of press, in respect to conventional progressive presses, is that the parts are transferred from one die to the next by means of so-called "fingers". Not only do the fingers transfer the parts but they also guide the component during the process. This allows parts to be drawn to the deepest depths with the tightest tolerances.

Other types of presses:

- Die-Set Transfer Press: Part is transferred via transfer fingers as the part progresses through the forming process. Tooling components attached to die plates enable the die to be installed in the press as one unit.

- ICOP (Individually Cam Operated Press): The part is transferred via transfer fingers as the part progresses through the forming process. Die components are installed in the press one station at a time.

- Progressive Die Press: The part is carried on the steel webbing as it progresses through the forming process.

Variations

Deep drawing has been classified into *conventional* and *unconventional* deep drawing. The main aim of any unconventional deep drawing process is to extend the formability limits of the process. Some of the unconventional processes include hydromechanical deep drawing, Hydroform process, Aquadraw process, Guerin process, Marform process and the hydraulic deep drawing process to name a few.

The Marform process, for example, operates using the principle of rubber pad forming techniques. Deep-recessed parts with either vertical or sloped walls can be formed. In this type of forming, the die rig employs a rubber pad as one tool half and a solid tool half, similar to the die in a conventional die set, to form a component into its final shape. Dies are made of cast light alloys and the rubber pad is 1.5-2 times thicker than the component to be formed. For Marforming, single-action presses are equipped with die cushions and blank holders. The blank is held against the rubber pad by a blank holder, through which a punch is acting as in conventional deep drawing. It is a double-acting apparatus: at first the ram slides down, then the blank holder moves: this feature allows it to perform deep drawings (30-40% transverse dimension) with no wrinkles.

Industrial uses of deep drawing processes include automotive body and structural parts, aircraft components, utensils and white goods. Complex parts are normally formed using progressive dies in a single forming press or by using a press line.

Workpiece Materials and Power Requirements

Softer materials are much easier to deform and therefore require less force to draw. The following is a table demonstrating the draw force to percent reduction of commonly used materials.

Drawing force required for various materials and reductions				
Material	Percent reduction			
	39%	43%	47%	50%
Aluminium	88	101	113	126
Brass	117	134	151	168
Cold-rolled steel	127	145	163	181
Stainless steel	166	190	214	238

Tool Materials

Punches and dies are typically made of tool steel, however cheaper (but softer) carbon steel is sometimes used in less severe applications. It is also common to see cemented carbides used where high wear and abrasive resistance is present. Alloy steels are normally used for the ejector system to kick the part out and in durable and heat resistant blankholders.

Lubrication and Cooling

Lubricants are used to reduce friction between the working material and the punch and die. They also aid in removing the part from the punch. Some examples of lubricants used in drawing operations are heavy-duty emulsions, phosphates, white lead, and wax films. Plastic films covering both sides of the part while used with a lubricant will leave the part with a fine surface.

PERMISSIONS

All chapters in this book are published with permission under the Creative Commons Attribution Share Alike License or equivalent. Every chapter published in this book has been scrutinized by our experts. Their significance has been extensively debated. The topics covered herein carry significant information for a comprehensive understanding. They may even be implemented as practical applications or may be referred to as a beginning point for further studies.

We would like to thank the editorial team for lending their expertise to make the book truly unique. They have played a crucial role in the development of this book. Without their invaluable contributions this book wouldn't have been possible. They have made vital efforts to compile up to date information on the varied aspects of this subject to make this book a valuable addition to the collection of many professionals and students.

This book was conceptualized with the vision of imparting up-to-date and integrated information in this field. To ensure the same, a matchless editorial board was set up. Every individual on the board went through rigorous rounds of assessment to prove their worth. After which they invested a large part of their time researching and compiling the most relevant data for our readers.

The editorial board has been involved in producing this book since its inception. They have spent rigorous hours researching and exploring the diverse topics which have resulted in the successful publishing of this book. They have passed on their knowledge of decades through this book. To expedite this challenging task, the publisher supported the team at every step. A small team of assistant editors was also appointed to further simplify the editing procedure and attain best results for the readers.

Apart from the editorial board, the designing team has also invested a significant amount of their time in understanding the subject and creating the most relevant covers. They scrutinized every image to scout for the most suitable representation of the subject and create an appropriate cover for the book.

The publishing team has been an ardent support to the editorial, designing and production team. Their endless efforts to recruit the best for this project, has resulted in the accomplishment of this book. They are a veteran in the field of academics and their pool of knowledge is as vast as their experience in printing. Their expertise and guidance has proved useful at every step. Their uncompromising quality standards have made this book an exceptional effort. Their encouragement from time to time has been an inspiration for everyone.

The publisher and the editorial board hope that this book will prove to be a valuable piece of knowledge for students, practitioners and scholars across the globe.

INDEX

www.ingramcontent.com/pod-product-compliance
Lightning Source LLC
Chambersburg PA
CBHW061243190326
41458CB00011B/3564